Making Prehistory

Scientists often make surprising claims about things that no one can observe. In physics, chemistry, and molecular biology, scientists can at least experiment on those unobservable entities, but what about researchers in fields such as paleobiology and geology who study prehistory, where no such experimentation is possible? Do scientists discover facts about the distant past or do they, in some sense, make prehistory? Derek Turner argues that this problem has surprising and important consequences for the scientific realism debate. His discussion covers some of the main positions in current philosophy of science – realism, social constructivism, empiricism, and the natural ontological attitude – and shows how they relate to issues in paleobiology and geology. His original and thought-provoking book will be of wide interest to philosophers and scientists alike.

DEREK TURNER is Assistant Professor of Philosophy at Connecticut College.

CAMBRIDGE STUDIES IN PHILOSOPHY AND BIOLOGY

General Editor
Michael Ruse   *Florida State University*

Advisory Board
Michael Donoghue   *Yale University*
Jean Gayon   *University of Paris*
Jonathan Hodge   *University of Leeds*
Jane Maienschein   *Arizona State University*
Jesús Mosterín   *Instituto de Filosofía (Spanish Research Council)*
Elliott Sober   *University of Wisconsin*

Recent Titles
Alfred I. Tauber   *The Immune Self: Theory or Metaphor?*
Elliott Sober   *From a Biological Point of View*
Robert Brandon   *Concepts and Methods in Evolutionary Biology*
Peter Godfrey-Smith   *Complexity and the Function of Mind in Nature*
William A. Rottschaefer   *The Biology and Psychology of Moral Agency*
Sahotra Sarkar   *Genetics and Reductionism*
Jean Gayon   *Darwinism's Struggle for Survival*
Jane Maienschein and Michael Ruse (eds.)   *Biology and the
Foundation of Ethics*
Jack Wilson   *Biological Individuality*
Richard Creath and Jane Maienschein (eds.)   *Biology and
Epistemology*
Alexander Rosenberg   *Darwinism in Philosophy, Social Science,
and Policy*
Peter Beurton, Raphael Falk, and Hans-Jörg Rheinberger (eds.)
The Concept of the Gene in Development and Evolution
David Hull   *Science and Selection*
James G. Lennox   *Aristotle's Philosophy of Biology*
Marc Ereshefsky   *The Poverty of the Linnaean Hierarchy*
Kim Sterelny   *The Evolution of Agency and Other Essays*
William S. Cooper   *The Evolution of Reason*

Making Prehistory

Historical Science and the Scientific Realism Debate

DEREK TURNER

Connecticut College

CAMBRIDGE
UNIVERSITY PRESS

CAMBRIDGE
UNIVERSITY PRESS

University Printing House, Cambridge CB2 8BS, United Kingdom

Cambridge University Press is part of the University of Cambridge.

It furthers the University's mission by disseminating knowledge in the pursuit of
education, learning and research at the highest international levels of excellence.

www.cambridge.org
Information on this title: www.cambridge.org/9780521875202

© Derek Turner 2007

First published 2007
First paperback edition 2012

A catalogue record for this publication is available from the British Library

ISBN 978-0-521-87520-2 Hardback
ISBN 978-1-107-40638-4 Paperback

For Michelle I. Turner

Contents

Contents

Figures

All figures are reprinted with permission of the Paleontological Society. Figures 1.1, 1.2, 1.3, 1.4 are from Wilson and Carrano (1999). Figure 4.1 is from Collins (1996).

Acknowledgments

In the spring of 2002, I presented some early ideas for this book at a philosophy of biology workshop at Florida State University. The thoughtful comments, criticism, and advice I received from the participants – André Ariew, Tim Lewens, Jason Robert, Betty Smocovitis, Kim Sterelny, and others – helped give shape to the project during the early stages. Most of all, I thank Michael Ruse, who hosted the workshop, for the kindness and generosity he showed an untested first-year assistant professor, and for his subsequent guidance and encouragement. During the early work on this project, Michael Lynch was a patient mentor and sounding board who helped me through several false starts, and this book owes a great deal to our frequent conversations over coffee in the Connecticut College student center. I am especially grateful to Todd Grantham and two anonymous readers for Cambridge University Press who read the entire manuscript with extraordinary care and provided me with invaluable suggestions and criticism. Audiences at meetings of the International Society for the History, Philosophy, and Social Studies of Biology (ISHPSSB) in Quinnipiac, CT, in July 2001, and Vienna, Austria, in July 2003, provided helpful feedback on material that would later make it into this book. Lauren Hartzell, Kate Kovenock, John Post, Brian Ribeiro, and Michelle Turner also read and commented on individual chapters or on papers that would later become chapters. My good friend Brian Ribeiro's work on skepticism has been a major influence. Students who took my philosophy of science courses at Connecticut College during the spring of 2004 and the spring of 2006 read portions of the manuscript and responded with tough and perceptive questions. Finally, I thank my colleagues at Connecticut College – Simon Feldman, Andrew Pessin, Kristin Pfefferkorn, Larry Vogel, and Mel Woody – for their confidence and their enthusiasm for the project, for their comments on work in progress, and for many wonderful conversations.

My work on this project was supported by two summer research stipends from Connecticut College (during the summers of 2002 and 2003) as well as a ConnSharp grant for summer research, with Kate Kovenock, during the summer of 2004.

Portions of the manuscript are based on material that has been published elsewhere. Chapter 2 appeared as "Local underdetermination in historical science," *Philosophy of Science* 72: 209–230, copyright 2005 by the Philosophy of Science Association, and reprinted with permission from the University of Chicago Press. Chapter 3 is a substantially reworked version of an earlier paper, reprinted from *Studies in History and Philosophy of Science* 35, "The past vs. the tiny: historical science and the abductive arguments for realism," pp. 1–17, copyright 2004, with permission of Elsevier. Chapter 4 is based in large part on a paper reprinted from *Studies in History and Philosophy of Science* 36, "Misleading observable analogues in paleontology," pp. 175–183, copyright 2005, with permission from Elsevier. I have also benefited greatly from critical comments provided by anonymous reviewers for these journals.

Introduction

Much of the sound and fury in the philosophy of science over the last few decades has had to do with a view – or better, a family of views – known as *scientific realism*. Pick up any issue of *Science* magazine, and you will find reports on research dealing with microphysical entities, properties, events and processes. For example, one article in the August 19, 2005 issue includes the following claim:

> When x-ray photons pass through a liquid sample that is thin compared to its x-ray absorption depth, less than 1% of the photons are scattered (Anfinrud and Schotte 2005, p. 1192).

Oversimplifying shamelessly (I will get to the finer distinctions later), the scientific realist thinks that scientists know a great many things like this, even though no one could possibly see or smell an x-ray photon, or bump into one while wandering about at night. The realist holds that a great many scientific claims like this one are true, or nearly true.[1] Those x-ray photons are really out there, and liquids really do have x-ray absorption depths – really! And what's more, scientists did not bring any of this about; they *discovered* it all. What happens to photons when they pass through a liquid sample does not depend at all on what we think about photons, or on the concepts we use to think about them, or on the language we use to talk about them. The history of science is a tale of progress in which scientists learn more and more (or get closer and closer to the truth) about how the world really is, independently of us. That's realism.

---

[1] What could it mean for a claim to be nearly, or approximately true? Realists have struggled to clarify the notions of approximate truth and verisimilitude, with mixed success. Indeed, the difficulty of explaining what approximate truth could be has driven many philosophers away from realism. See Psillos (1999, ch. 11) for one helpful recent discussion of this issue from a realist perspective.

But x-ray photons are one thing; dinosaurs, shifting tectonic plates, and evolutionary processes also pose a challenge. Should we be realists about those things? Should we be realists about prehistory?

Most of the philosophers who think and write about scientific realism take their examples from the study of the microphysical world. Sadly, historical sciences, such as paleobiology and geology, have been left almost entirely out of the discussion, even though one cannot see, or smell, or bump into a living dinosaur any more than one can an x-ray photon.[2] As a result, I argue, the scientific realism debate has been skewed. I have written this book with two audiences in mind: First, I hope to show philosophers of science how our assessment of the arguments for and against scientific realism, and of some of the main positions that philosophers have staked out in the realism debate, might change when we examine them with an eye toward the scientific study of prehistory. Second, I hope to show scientists who study prehistory that the scientific realism debate, contrary to the impression one would get from perusing the philosophical literature, has relevance to their work, and may even have the potential to change the way they conceive of what they do.

One tried and true recipe for a philosophical book is to begin with one or two claims that strike everyone as boring, obvious, and uncontroversial; then show, by a series of unimpeachable logical steps, that these claims have surprising, counterintuitive consequences whose truth no one ever would have suspected. The more boring and uncontroversial the original claims, and the more surprising and counterintuitive the consequences, the better.

In this book, I begin with two boring and obvious claims about how the past differs from the microphysical world. I give these two claims high-sounding names – the *asymmetry of manipulability* and the *role asymmetry of background theories* – but there is nothing fancy or even very subtle about the ideas themselves. The first idea is that although we obviously cannot change the past, we can use technology to manipulate things and events at the microphysical level. Scientists have designed particle accelerators that make it possible for them to run experiments in which they crash subatomic particles together. For other vivid examples of technological control of microphysical events and processes, think of nuclear

---

[2] See, however, Wylie (2002, ch. 5) for a defense of scientific realism in the context of archaeology. Wylie just presents the case for realism in general and concludes that we should be realists about the past. She does not consider the possibility that the strength of the arguments for and against realism might vary depending on the scientific context.

weapons, or genetic engineering, or current research in nanotechnology. The second idea has to do with what philosophers of science call background theories, or theories that scientists take for granted in the course of their research. In historical science, background theories all too often tell us how historical processes destroy evidence over time, almost like a criminal removing potential clues from a crime scene. For example, the fossilization process destroys all sorts of evidence about the past, with the result that we will never know many things about the past, such as the colors of the dinosaurs. In experimental science, by contrast, background theories more often suggest ways of creating new empirical evidence. For example, one can scarcely begin to understand the development of modern physics and astronomy without appreciating how the study of optics, or the behavior of light as it passes through lenses, bounces off mirrors, and so on, contributed to (and also benefited from) the development of ever more sophisticated microscopes and telescopes. More generally, part of the point of experimentation in science is to create new evidence, and background theories about microphysical entities and processes often suggest new ways of doing that. Taking quantum theory for granted enables scientists to build particle accelerators, which in turn enable them to run new kinds of experiments. In historical science, background theories often tell scientists how the evidence has been destroyed; in experimental science, they often tell scientists how to manufacture new evidence.

Hopefully all of this sounds like common sense. In this book, I undertake to show that these fairly obvious ideas have important and surprising consequences that most philosophers of science have yet to appreciate. In the first part of the book (chapters 1 through 5), I examine the main arguments for and against scientific realism, and I show that the strength of those arguments varies in interesting and sometimes complicated ways, depending on whether we are talking about the microphysical world or about prehistory. For example, chapter 5 argues that novel predictive successes will be fewer and further between in historical than in experimental science. If that is right, it bears directly on one of the most popular current arguments for scientific realism: the argument that interpreting theories realistically is the best way to make sense of their novel predictive successes. This is one of the things I mean by saying that the scientific realism debate has been skewed by the neglect of geology and paleobiology. I also look at the consequences that the asymmetry of manipulability and the role asymmetry of background theories have for the underdetermination problem (chapter 2), the more traditional arguments for scientific realism (chapter 3), and the pessimistic induction from the history of science (chapter 4).

The title of this book, *Making Prehistory*, hints at the sort of social constructivist views that many scientists find kooky, or worse. You may be thinking: "Surely he's not going to argue that dinosaurs are social constructs, or that their extinction is something that the scientific community – somehow – brought about." Don't worry; I am not going to argue that. But I am not a scientific realist, either, at least not across the board. Instead, I defend a view, the *natural historical attitude*, which is inspired by the work of the philosopher Arthur Fine (1984, 1986, 1996). Fine, caring more about physics than about geology or paleobiology, called his own view the "natural ontological attitude." The natural historical attitude is one of agnosticism with respect to the metaphysics of the past: Maybe we have made prehistory, and maybe we haven't. But if we take our own best theories about the past seriously, they clearly imply that we will never have any historical evidence that could adjudicate the dispute between metaphysical realists and social constructivists, so we do best to suspend judgment and move on to other things. That is the take-home message of chapter 6.

Among philosophers of science, the most respectable alternatives to scientific realism are Arthur Fine's natural ontological attitude and Bas van Fraassen's constructive empiricism, a radical view which has it that our knowledge is entirely restricted to what we can observe. Van Fraassen, like Fine, has concerned himself mainly with physics, and both of these versions of non-realism look like genuine contenders so long as we restrict our attention to the microphysical world. However, I argue in chapter 7 that when we turn our attention to the scientific study of prehistory, van Fraassen's view has such repugnant consequences that it must drop out of serious contention. This, incidentally, is another way in which the realism debate has been skewed by the neglect of historical science: Van Fraassen's constructive empiricism seems at first like a viable philosophical theory of science, but only so long as we ignore geology and paleobiology.

In the concluding chapter, I take up the issue of consilience, or the idea that scientists can have some confidence that they are getting things right when they can offer a unified explanation of a variety of seemingly unrelated phenomena. What should someone who takes seriously the asymmetry of manipulability and the role asymmetry of background theories say about consilience? How might our understanding of the role of consilience in historical science be affected by adopting the natural historical attitude? I argue that while appeals to consilience do have some evidential weight, the asymmetries also mean that scientists should be moderately skeptical about such appeals.

Most scientists who work at reconstructing the past seem to take a realist view of prehistory. This consensus, or near consensus, can make it seem as though realism were the most natural or most obvious position. One potential explanation for this near consensus is that philosophers of science have not articulated any serious non-realist alternatives. Another potential explanation is that none of the great theories of historical science – evolutionary theory, plate tectonics, etc. – cause trouble for scientific realism in quite the way that quantum theory does.

During the twentieth century, the scientific realism debate evolved in step with significant changes in theoretical physics. Disagreements about how to interpret quantum theory, for example, became tangled up with disagreements about whether to adopt a realist or an instrumentalist interpretation of scientific theories. Without going into details, we can note that quantum theory has two features which, taken together, raise some pretty basic philosophical questions: First, that theory has proven itself to be wildly successful at generating accurate predictions. And second, if we take literally what quantum theory implies about the microphysical world – for example, about the superposition of states, about the collapse of the wave function, about non-locality, and much else – the theory seems wildly unfamiliar and counterintuitive. These two features have driven many philosophers and scientists towards *instrumentalism*, or the view that scientific theories are just instruments or tools for generating predictions. Instrumentalism treats scientific theories as a kind of technology. If quantum theory is merely a complex mathematical tool for generating accurate predictions, in exactly the same way that a hammer is a tool for driving nails, then there is no need even to ask whether the theory accurately represents the microphysical world. Contrary to realists, instrumentalists hold that truth and accurate representation are not what science is all about. Instead, science is all about results, and about increasing our control of the world around us. At any rate, since theories in physics often naturally and inevitably give rise to the sorts of questions that animate the realism debate, probably no one needs to explain why physicists should care about that debate. But what about geologists and paleobiologists? What might they stand to gain from an exploration of the realism debate?

Although this book is mainly an essay on scientific knowledge, many of the questions raised here also have to do with issues of status and prestige. Within biology, for example, cell and molecular biology and anything involving medical research tend to enjoy a somewhat higher status. Subfields such as ecology, and anything involving whole organisms, enjoy a somewhat lower status. At the low end of the totem pole, we find paleontologists, who study whole organisms that do not even exist anymore. In his classic, *Wonderful Life* (1989), Stephen Jay Gould makes an impassioned "plea for the high status of natural history" and laments the fact that people so often associate the experimental method with *the* scientific method. He quotes the physicist Luis Alvarez – ironically, one of the formulators of the hypothesis that an asteroid collision caused the mass extinction at the end of the Cretaceous period – as saying: "I don't like to say bad things about paleontologists, but they're really not very good scientists. They're more like stamp collectors" (1989, p. 281). Of course, it is not true that all paleontologists do is to collect specimens from the field and publish descriptions of them, but the quotation reveals something about how people have perceived the study of prehistory. Gould, for his part, argues with great passion and eloquence for a view that could be summed up by the slogan, *Different Methods, Epistemic Equality*. That is to say, historical science and experimental science necessarily employ different methods of investigation, as well as different styles of explanation, and they emphasize different things (particulars *vs.* laws and regularities). But according to Gould and some other more recent writers (such as Carol Cleland, whose work I discuss in chapter 2), these methodological differences make no significant epistemological difference: When it comes to delivering scientific knowledge, historical work is every bit as good as experimental work.

I would like to renew Gould's plea for the high status of historical science. However, Gould goes about making that plea in a counterproductive way. The asymmetry of manipulability and the role asymmetry of background theories really do place historical researchers at a relative epistemic disadvantage, so the slogan "Different Methods, Epistemic Equality," is mistaken. In its place I would propose a different slogan: *Epistemic Disadvantage, Equal Scientific Status*. I try to drive this point home in chapters 2 through 5 by examining the main arguments that philosophers of science have discussed in connection with the realism debate. In my view, rather than denying the epistemic disadvantages of historical science, we can make the best case for the high status of natural history by calling attention to those disadvantages and even celebrating them. If we were watching two distance runners, one of whom runs along a smooth track (perhaps even one that is outfitted with one of those moving walkways you find in airports), while the other runs along hilly and treacherous terrain, we should think very highly of the second runner, even if she takes longer to cover the same distance. Acknowledging that those who study the past find themselves at an epistemic disadvantage relative to those who study the microphysical world is also the key to understanding some of the most interesting developments in paleobiology and geology over the last few decades, such as the use of computer simulations to carry out numerical experiments. Numerical modelling is a strategy for coping with the asymmetry of manipulability.[3]

What else might we gain from this exploration of the consequences of the two asymmetries, and of the ways in which the scientific realism debate has been skewed toward the microphysical? For too long, discussions of historical science have been constrained by the traditional distinction between *ideographic* and *nomothetic* science.[4] We owe that terminology to the neo-Kantian philosophers, Wilhelm Windelband and Heinrich Rickert, who thought that this distinction shed some light on the differences between the natural sciences (*Naturwissenschaften*) and the human sciences (or *Geisteswissenschaften*). According to this tradition, nomothetic science is concerned with laws and regularities, or with patterns involving types of events. Ideographic science, by contrast, focuses on sequences of particular events, or on event tokens. Ideographic science, not surprisingly, is often thought to involve some sort of narrative. Kepler and Newton were doing nomothetic science. The nineteenth-century geologists who first drew the inference that much of the northern hemisphere was once covered by an ice pack were doing ideographic science. For my part, I have not found the ideographic/nomothetic distinction to be very helpful. Paleontologists have taken advantage of laws of biomechanics to infer how fast a dinosaur was walking when it made a particular set of tracks (Alexander 1976). It is also possible to use biomechanical considerations to deduce the maximum swimming speeds of extinct marine reptiles (Massare 1988). Geologists run elaborate computer simulations

---

[3] As will become apparent, my interpretation of these recent developments is deeply influenced by Huss (2004).

[4] For a helpful discussion of this distinction, see Tucker (2004, p. 241). I have also found Stephen Jay Gould's (1987) to be very illuminating. Much of Gould's work in the 1970s and 1980s was animated by the idea that paleobiology need not be an entirely ideographic discipline. According to Gould, paleontology "resides in the middle of a continuum stretching from idiographic to nomothetic disciplines" (1980, p. 116). Gould, Raup, Sepkoski, and Simberloff began using stochastic models of evolution during the 1970s, and they saw that as an attempt to make paleobiology into a more nomothetic discipline (see Raup et al. 1973; Raup and Gould 1974).

to test ideas about what the earth's climate might have been like 600 million years ago. Each simulation models a series of particular events, but scientists run the models over and over again, refining them and adjusting parameters as they go. Are these examples of ideographic or of nomothetic science? What could we gain by forcing these examples into one category or the other? I aim to show that we can get a much more realistic picture of historical science (in the ordinary, not the philosophical sense of "realistic") if we cut loose from this distinction between ideographic and nomothetic science and focus instead on the epistemically relevant differences between the different kinds of unobservable things that scientists study.

Finally, why should scientists care about the natural historical attitude? For I really do recommend that attitude as a good one for geologists, paleobiologists, and even archaeologists and historians to adopt. But what difference would such an attitude make to working scientists? I offer two answers to this question. First, the disconnect between philosophical discussions of the arguments for and against scientific realism, on the one hand, and historical science, on the other, has left scientific realism as the default view of the sciences of prehistory. Since no one has articulated any serious alternatives to realism with respect to geology and paleobiology, realism is the only game in town. The few philosophers who have thought deeply about non-realist views about the past – for example Michael Dummett – have had little or no interest in the details of the practice of historical science. My worry is that when a certain philosophical view seems to be the only game in town, there is not much incentive for anyone to enter into an open and critical discussion of the fine points of that view. At any rate, I will argue that certain parts of the realist view of the past – especially the part about the past having occurred independently of us – go way beyond what is justified by the historical evidence. And I recommend the natural historical attitude as a stance that is more Spartan and less burdened with philosophical theory than metaphysical realism, and one that evinces greater respect for the limitations of historical evidence.

Second, over the last few decades, scientists and philosophers alike have been caught up in what have come to be known as "the science wars" (for a wonderful discussion, see Parsons 2001). This cultural conflict has pitted scientists and a great many professional philosophers of science (on all sides of the realism debate) against a variety of social constructivists, historians, and social theorists of science. In many ways, this conflict has been a clash of disciplinary methods and standards, but it has also involved substantive philosophical claims about the world, with one side claiming that scientists discover facts about the world, and the other saying that the scientific community constructs those facts. Perhaps the biggest consideration in favor of the natural historical attitude – and a consideration that I hope will appeal to scientists as well as to philosophers and social theorists of science – is that by adopting it we can put the so-called "science wars" behind us for good. Those who adopt the natural historical attitude can look back on the science wars as a dispute between two parties, both of whom were irrationally wedded to metaphysical claims that went beyond what the available evidence could possibly support. Not that the metaphysical realism/social constructivism issue is the only thing at stake in the science wars, but it is one of the most important things. I recommend the agnostic stance of the natural historical attitude as a compromise position.

1

Asymmetries

Several of the natural sciences – geology, paleontology, evolutionary biology, cosmology, and archaeology – purport to give us knowledge of prehistory. By "prehistory" I just mean everything that happened before the
invention of writing made it possible for people to leave written testimony for later investigators. This book is about those sciences, though it
deals mainly with the quintessentially historical sciences of paleontology
and geology. There are limitations and obstructions to our knowledge of
prehistory that do not similarly constrain our knowledge of the present
microphysical world. Putting it very roughly for now, this means that there
is a sense in which we can know more about the tiny than we can know
about the past. This is an example of an *epistemic asymmetry*, or lop-
sidedness in our scientific knowledge. In this opening chapter, I present
and explain the sources of this asymmetry. I then go on to indicate why
I think this asymmetry is so important, and why philosophers, scientists,
and indeed anyone with an interest in the scientific study of the past, ought
to care about it.

I begin by attempting to convince you that this epistemic asymmetry
between the past and the tiny is real, and that it is something we must
contend with. Next, I will show how we might go about explaining this
epistemic asymmetry in terms of two deeper asymmetries, which I will call
the *asymmetry of manipulability*, and the *role asymmetry of background
theories*. The rest of the book draws out the surprising consequences of
these deeper asymmetries.

I.I LIMITS TO OUR KNOWLEDGE OF PREHISTORY

We can begin with an example of recent paleontological work in which
scientists seem to bump up against the limits to our knowledge of the

10

A   B

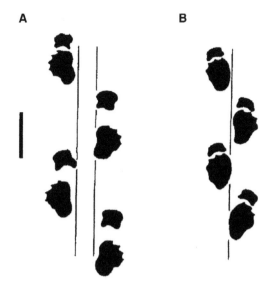

Figure 1.1   Wide-gauge (A) vs. narrow-gauge sauropod trackways. The bar on the left drawn for scale represents 1 meter.

distant past.[1] Scientists who work on problems in geology and paleontology often experience these limits in the following way: Everything up to a certain point has the feel of good, solid research. But everything beyond that point has the feel of speculation, educated guesswork, or (at worst) mere storytelling. Many scientists go ahead and cross this boundary once in awhile, although they usually find ways of signaling their awareness that they have crossed it. I have chosen the following example because it is one in which the boundary crossing is particularly vivid. It represents just one of several kinds of work that paleobiologists do. I will describe the example in some detail – more detail than philosophers usually allow themselves – because I want to convey what it feels like to make great progress in historical science before suddenly getting stymied.

The sauropod dinosaurs were the big, long-necked and relatively small-brained plant-eaters, such as *Brontosaurus* (a.k.a. *Apatosaurus*). The trackways of sauropod dinosaurs come in two basic varieties, known as wide-gauge and narrow-gauge, as depicted in figure 1.1. The main difference between these two concerns the distance of the footprints from the

---

[1] For another interesting discussion of limits to our knowledge of the past, see Tucker (2004, ch. 6). See also Tucker (1998) on the importance of the uniqueness of historical events.

midline of the track. In some narrow-gauge tracks, the left and right feet landed right on the midline. In wide-gauge tracks, the left and right feet were planted some distance from the midline. Most of the sauropod tracks dating from the Jurassic period (195 to 140 million years ago) are narrow-gauge. Wide-gauge tracks begin to show up in late Jurassic rocks, and most of the sauropod tracks from the early Cretaceous are wide-gauge. Which sauropods made which tracks? This is just one instance of a very general problem in paleontology – namely, figuring out how to match the available skeletal remains with other fossilized traces, or ichnological evidence.

One possibility is that the wide-gauge and narrow-gauge tracks were made by members of the same species. Perhaps juveniles made the narrow-gauge tracks while adults made the wide-gauge tracks. Or perhaps the two kinds of trackways represent two different gaits or walking styles. Maybe the sauropods walked with their legs spread apart, but ran with their legs directly beneath their bodies. Unfortunately, neither of these hypotheses has much plausibility. First, the footprints left by wide-gauge and narrow-gauge trackmakers are, on average, about the same size, which rules out the hypothesis that the wide-gauge trackmakers were just grown-up animals. Second, we know that the bigger an animal gets, the greater the biomechanical stresses on its legs, and the more difficult it is to change from one gait to another. Small mammals change gaits frequently and easily, while larger animals, such as elephants and rhinoceroses, are more restricted by their size. Elephants, for example, cannot gallop. The sauropods in question were much larger than elephants, and the hypothesis that a single animal was capable of making either wide- or narrow-gauge tracks is biomechanically implausible. Yet a third possibility is that the difference between wide-gauge and narrow-gauge tracks had to do with the substrate the animals were walking on. Perhaps the animals spread their legs wider when walking through sand or mud. But scientists have checked this, and they have found no correlation between track type and substrate type. That leaves the hypothesis that different types of sauropods made different types of tracks.

The two most plausible candidates for the wide-gauge tracks are the brachiosaurs (including *Brachiosaurus*, which probably weighed at least 80 or 90 tons) and the somewhat smaller, but still humongous, titanosaurs. The hypothesis that titanosaurs made the wide-gauge tracks gets a little extra support from the observation that most of the tracks occurring in rocks from the Cretaceous period, when titanosaurs flourished, are

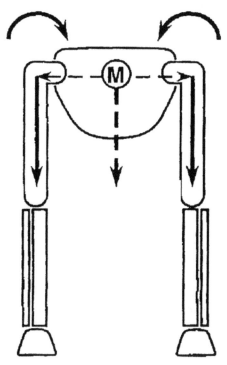

Figure 1.2 Mediolateral and compressive forces. This diagram represents the forces that act on an animal's limbs when its center of mass (M) is suspended between them. Since the limbs are not directly beneath the center of mass, they are subject to mediolateral, or bending forces (represented by the curved arrows) as well as compressive forces (represented by the straight arrows).

wide-gauge. Two paleontologists, Jeffrey Wilson and Matthew Carrano (1999), give an additional biomechanical argument to clinch the case for the titanosaurs as wide-gauge trackmakers. After giving this rigorous biomechanical argument, they self-consciously proceed to cross the boundary that separates solid science from speculation.

The legs of any large quadruped are subject to two kinds of forces, as shown in figure 1.2. The first is a compressive force that results from the fact that the legs must support the animal's mass. The second is a mediolateral, or bending force that is due to the fact that the animal's hip joints are some distance away from its center of mass. At this point, generalizations of biomechanics come into play. It so happens that there are three ways in which to increase the bending force that is exerted upon an animal's femora. The bending force increases, first, as the animal's body mass increases; second, as the hip joints move further apart; and third, as the left and right feet move further apart. This applies to humans as well: a person in a standing position can increase the bending force exerted upon his legs simply by spreading his feet apart. These biomechanical generalizations can all be confirmed by observation of living organisms. Wilson and Carrano then use these generalizations to infer that the leg bones of the wide-gauge trackmakers would have to be able to withstand greater mediolateral forces. Thus, one should expect the femora of the wide-gauge trackmakers to be thicker along the mediolateral axis.

The hindlimb bones of the sauropod dinosaurs exhibit just the sort of morphological variation that one would expect to see, given the hypothesis that the titanosaurs made the wide-gauge tracks, together with the biomechanical assumption that the femora of the wide-gauge trackmakers would have to withstand greater mediolateral stress. As shown in figures 1.3 and 1.4, the femur of *Diplodocus* has a straight shaft. An axis drawn from one condyle to another (that is, between the two ends of the bone) intersects the horizontal axis at a right angle. On the other hand, the femur of *Titanosaurus* has a condyle-to-condyle axis that intersects the horizontal axis at an angle greater than 90°, and a cross-sectional view of the *Titanosaurus* femur shows that it has a larger diameter than that of *Diplodocus*. This biomechanical line of reasoning leads unambiguously to the conclusion that titanosaurs made wide-gauge tracks, while brachiosaurs and diplodocids probably made narrow-gauge tracks.

So far, so good. But at this point, it is hard not to wonder why the titanosaurs, but not the brachiosaurs or diplodocids, made wide-gauge tracks. Why were these animals built differently? What, if anything, was the wide-gauge stance *for*?[2] Did the wide-gauge stance confer some sort of selective advantage? Wilson and Carrano suggest that their work lends some support to the hypothesis that the titanosaurs were semi-bipedal. Like the extinct giant ground sloths of much more recent times, they might have reared up on their hind legs to reach vegetation growing high above the ground. Wilson and Carrano list a number of anatomical features in saltasaurids (one group of titanosaurs) that add support to this hypothesis:

> These features include vertebral adaptations for increased trunk and tail mobility, changes in knee and elbow morphology resulting in greater

---

[2] This "What for?" question has a strong teleological flavor. The literature on this subject is extensive. See, for example, the anthologies edited by Buller (1999) and Allen, Bekoff, and Lauder (1998), as well as Turner (2000).

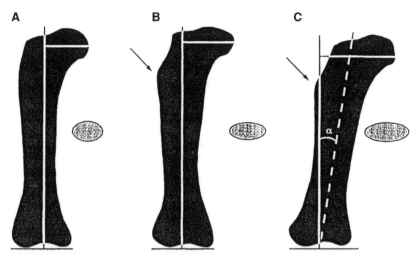

Figure 1.3   Femora of three sauropod dinosaurs: (A) *Diplodocus*, (B) *Brachiosaurus*, and (C) *Saltasaurus*, which is one of the titanosaurs. Note that (A) and (B) have straight shafts, whereas in (C), the horizontal axis intersects the condyle-to-condyle axis (represented by the dotted line) at an angle greater than 90°. Note also that the cross-section of (C) is more elliptical, suggesting that (C) was better able to withstand mediolateral stress.

flexibility, and wider foot stances for greater stability of the wider body carriage. More routine use of bipedal posture in saltasaurids is suggested by flared ilia for support of the viscera and by other features. (Wilson and Carrano 1999, p. 265)

But the scientists advance this hypothesis with great caution – so much caution, in fact, that it is hard to tell if they really mean to advance it at all:

These features are not proof of bipedalism in saltasaurids, and bipedalism is not required to explain their presence. No single feature even implies this behavior. Taken as a whole, however, saltasaurid (and other titanosaur) postcranial morphology strongly suggests that these sauropods exhibited distinct locomotor specializations relative to other sauropod groups. (Wilson and Carrano 1999, p. 265)

One thing that the hypothesis has going for it is *consilience*, which is a theme of chapter 8: It unifies, makes sense of, and pulls together a number of otherwise puzzling anatomical features. Is consilience enough?

A   B

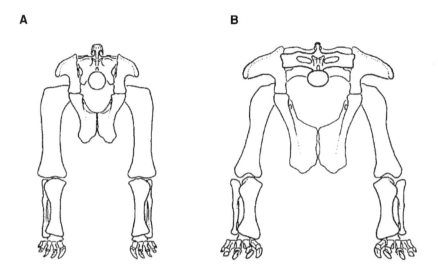

Figure 1.4 Hindlimb morphology of two sauropod dinosaurs: (A) *Camarasaurus*, and (B) *Opisthocoelocaudia*, one of the titanosaurs.

The conclusion that the titanosaurs made the wide-gauge tracks seems forced by the morphological evidence. It would be surprising, to say the least, if the brachiosaurs, whose legs were not built to withstand added mediolateral stress, had made the wide-gauge tracks, while titanosaurs made the narrow-gauge tracks. But the conclusion that the titanosaurs were semi-bipedal does not seem forced at all. With a little imagination, we could dream up some other account of the evolution of the wide-gauge stance – perhaps it had something to do with mating behavior, rather than foraging. Moreover, it is hard to see what further tests scientists could use to determine whether the animals were, in fact, semi-bipedal, for we can never observe titanosaurs in action.

We can reasonably claim to know that titanosaurs made the wide-gauge tracks. But we cannot so reasonably claim to know that the titanosaurs were semi-bipedal. What this case illustrates is that in attempting to recon-struct the distant past, scientists can only go so far. At a certain point, researchers are bound to pass from good, solid science, to explanatory speculation and educated guesswork.

In this book, I will not defend a general theory of scientific knowledge, or try to draw the limits to our knowledge of prehistory with precision. Instead, I want to try to understand why there are any such limits at all, and why our knowledge of the microphysical world is not limited in the same way, or to the same extent.

I.2 THE TIME ASYMMETRY OF KNOWLEDGE

In order to set the stage for the thesis that there is an epistemic asymmetry between our knowledge of the past and our knowledge of the tiny, it will help to begin by considering a far less controversial example of an epistemic asymmetry. (The term "epistemic" comes from *episteme*, the Greek word for knowledge.) The time asymmetry of knowledge will serve well as an analogue for the thesis I shall defend.

We appear to know quite a bit more about the past than we do about the future. Anyone can recall what the weather was like yesterday or the day before, and we can consult the meteorological records to learn what the weather was like on this day one year ago. However, it is difficult enough to predict what the weather will be like tomorrow, and no one can reasonably claim to know what the weather will be like one year from today. This difference between the past and the future also applies to human affairs. For instance, we all know who won the US presidential election in 2004, but nobody knows who will win in 2008. Each of us knows when and where we were born, but not when and where we will die. Thus, our knowledge seems lopsided; it seems easier to know things about the past than to know things about the future, but why? Intuitively, it seems that there must be something that limits or obstructs our knowledge of the future, but not our knowledge of the past. We can call this idea the *time asymmetry of knowledge.*

If knowledge were really time asymmetrical, that might help make sense of the fact that while a number of respectable disciplines are devoted to the study of the past – not only paleontology and geology, but also archaeology, history, evolutionary biology, historical linguistics, and so forth – there is no such science as futurology.

One might think that the explanation of the time asymmetry of knowledge is just obvious: Future events have not yet occurred, and that explains why it is more difficult to acquire knowledge of the future than of the past. While this sounds right, it amounts to little more than a restatement of the phenomenon to be explained. Since it is true by definition that future events have not yet occurred, saying that it is difficult to acquire knowledge of the future because future events have not yet occurred is like saying that it is difficult to acquire knowledge of the future because it is the future.

17

Paul Horwich (1987) offers a different explanation of the time asymmetry of knowledge.[3] He argues that while there are *recording systems* that provide us with information about the past, there are no analogous *precording systems* that would provide us with information about the future.[4] The absence of precording systems is what obstructs our knowledge of the future. Or equivalently, the relative abundance of recording systems makes it possible for us to know a great deal more about the past.

To begin with, Horwich gives an abstract account of an ideal recording system.[5] Probably no recording systems are ideal, but all recording systems, from photographs to fossilized trackways, approximate this ideal to one degree or another. According to Horwich, an ideal recording system, S, has three essential features:

1. S is capable of being in any of a range of mutually exclusive states, $S0$, $S1$, $S2$, . . .
2. Except for $S0$, these states are perfectly stable; that is, if S is in state Sk at time t, then S is in state Sk at all times later than t.
3. There exists a range of mutually exclusive external conditions $C1$, $C2$, . . ., to which S is sensitive in the following sense: if S is in its "neutral" state $S0$ at time t, and the external condition Ck obtains in the environment of S, then S will go immediately into state Sk; moreover this is the only way that Sk can be produced. (Horwich 1987, p. 84)

Think of a sandbox as a simple recording system. To begin with, the sandbox is in the neutral state $S0$: someone has smoothed out the sand with a rake. If the causal condition, Ck, obtains – say, if Cory walks through the sandbox – the system will go into state Sk, which is to say that there will now be a set of footprints. The sandbox is not an ideal recording system because it does not have the second of the above features. Some children could come along and disturb the tracks, so there is no guarantee that the system will stay in state Sk forever. In addition, condition Ck might not

---

[3] Horwich (1987, ch. 5) entertains and ultimately dismisses a number of intriguing potential explanations of the time asymmetry of knowledge. He rejects the strategy of explaining the time asymmetry of knowledge in terms of the time asymmetry of overdetermination (pp. 81–82). Horwich's argument on this score accords with the argument I offer later on in ch. 2.

[4] Horwich actually uses the term "pre-recording."

[5] For a related account of the nature of recording devices, see Feinberg, Lavine, and Albert (1992, pp. 635–637).

be the only thing that could drive the system into state *Sk*. If Elmer and Cory wear the same shoe size, then Elmer could also make the system go into state *Sk* by walking through the sandbox. Although the sandbox is far from an ideal recording device, it still conveys information about the past.

But what would a *precording system*, even a less than ideal one, look like? A precording system, S^*, must have the following features:

1. S^* is capable of being in any of a range of mutually exclusive states $S0$, $S1, S2, \ldots$

2. Except for $S0$, these states are fairly stable; that is, if S^* is in a state *Sk* at time t, then S^* is probably in state *Sk* at all times *earlier* than t.

3. There exists a range of mutually exclusive external conditions $E1, E2, \ldots$, with which S^* is associated in the following way: if S^* is in its neutral state at time t, and the external condition Ek obtains in the environment of S^*, then *beforehand*, S^* was in state *Sk*; moreover this is what usually happens following *Sk*. (Horwich 1987, p. 87)

The sandbox is obviously not a precording system in this sense. Suppose, as before, that Cory walks through the sandbox at time t. His walking through the sandbox is the effect condition (Ek), and the tracks are state *Sk*. If the sandbox were a precording device, then it would have to have been in state *Sk* even *before* Cory walked through it. In other words, the tracks would have to precede Cory's walking through the sand. More generally, if the sandbox were a precording device, its earlier states would have to convey information about later effect conditions, but sandboxes do not work this way. They can serve as records of what happened in the past, but they cannot serve as precords of what will happen in the future. Indeed, Horwich generalizes this point: We have lots of available records, but relatively few precords. The absence of precording systems severely limits our knowledge of the future, but not our knowledge of the past, and that explains why our knowledge seems so lopsided with respect to time. We have, for example, the fossil and the geological records, but no similar precords of the distant future.[6]

---

[6] One might think that if determinism were true, then the entire state of the universe at a time could be thought of as an ideal precording system. Laplace's demon, for example, could deduce all the facts about the future from the facts about the present conjoined with the laws of nature. Strictly speaking, though, the state of the universe at a time would not be a precording system, in Horwich's sense, because a precording system is a system whose states are correlated with certain external conditions. In order to treat the entire universe

In addition to pointing out that our world contains lots of recording systems but few precording systems, there might be another way of explaining the time asymmetry of knowledge. Horwich also thinks that the epistemic asymmetry between the past and the future has to do with something known as the *fork asymmetry*, which he characterizes as follows:

> [G]iven a strong correlation between events *A* and *B*, there is always some explanation – some earlier event *C* – that causes them both. This fact is time-asymmetric, for it is frequently not the case that correlated events *A* and *B* have a characteristic joint effect *E*. (Horwich 1987, p. 73)

Correlated events usually, if not always, have common causes, but seldom, if ever, have common effects.

Most philosophers of science use the resources of probability theory to help explicate the notions of correlation and common cause. For example, we can say that there is a positive correlation between two events, A and B, when

$$\text{Prob(A and B)} > \text{Prob(A)} \times \text{Prob(B)}$$

In other words, the probability of both A and B occurring together is greater than the product of the independent probabilities of A and B. Moreover, we can say that C is the common cause of A and B when it screens off A from B, in the following sense:

$$\text{Prob(A} \mid \text{B and C)} = \text{Prob(A} \mid \text{C)}$$

In other words, the probability of A given B and C equals the probability of A given C alone. (The common cause, C, also screens B from A in the same way.) Consider, by way of example, the clues left at a crime scene: a shattered car window, and a hole in the dashboard where the stereo once resided. The probability of these two things occurring together is greater than the product of their independent probabilities. That is, the probability that the window is shattered and the stereo missing

---

as a precording or recording system, there would have to be external conditions that its states are correlated with. Horwich's point is that *even if our universe is deterministic*, there is an asymmetry of recording and precording devices. I thank Andrew Pessin for calling my attention to this issue.

is greater than the probability that the window is shattered, times the probability that the stereo is missing. The common cause of these two clues is the thief's breaking into the car and stealing the stereo. What this means is that the probability that the stereo is missing, given that the window is broken and that the thief broke in and stole it, equals the probability that the stereo is missing, given only that the thief broke in and stole it. This example also helps to illustrate the fork asymmetry. Correlations (such as that between missing car stereos and broken car windows) usually have common causes. However, they seldom have common effects.

To return to the paleontological example, we can think of the scientists as positing a common cause of the footprints and skeletal remains in the fossil record. The titanosaurs were the common cause of both the skeletons and the tracks, just as the thief is the common cause of the broken window and the missing stereo. The conclusion that the skeletal remains and the fossilized trackways have a common cause may seem so obvious as to be barely worth discussing, but it illustrates an important inference pattern. Indeed, the logical empiricist philosopher Hans Reichenbach famously advanced the so-called *principle of the common cause* as a methodological principle in science. According to him, scientists should always try to explain correlations by positing common causes. More recently, a number of authors have suggested that historical science proceeds by positing common causes of historical traces (Sober 1988; Cleland 2002; Tucker 2004).

If the thesis of the fork asymmetry is correct, it might also explain how we are (in a sense) able to know more about the past than about the future. The fact that correlated events typically do not have common effects imposes a limit on our knowledge of the future by making it difficult to draw future-oriented causal inferences from correlations. We can infer that the fossilized footprints and skeletal remains have a common cause (or more precisely, causes of a common type), but we have no reason at all to think that they have any common effect, and so we cannot use this correlation to draw any conclusions about the future. Of course, we can still know a lot about the future; Horwich's suggestion is only that the fork asymmetry makes it relatively easier to acquire knowledge of the past.

Horwich himself argues that the fork asymmetry affords the deeper explanation of the time asymmetry of knowledge, because the fork asymmetry explains why there are recording systems but not precording systems.[7] Moreover, he tries to offer an even deeper explanation of the fork asymmetry.[8] For present purposes, we need not worry about these aspects of his argument. All I want to suggest here is, first, that knowledge exhibits a time asymmetry, and second, that philosophers have gone some way toward explaining why knowledge is time asymmetrical. In general, the way to explain an asymmetry is by tracing it to a deeper, more fundamental asymmetry. In this case, we have traced the time asymmetry of knowledge to the asymmetry of recording and precording devices, and to the fork asymmetry.

Before moving on, two more observations concerning the time asymmetry of knowledge seem relevant. First, the claim that knowledge exhibits a time asymmetry does not tell us, by itself, how much we know (or can know) about the future. Nor does it tell us how much we know (or can know) about the past. The thesis that knowledge is time asymmetrical is compatible, first of all, with our knowing a great deal about the future. It could turn out that we have vast amounts of knowledge of both the past and the future, even though our knowledge is time asymmetrical. On the other hand, the thesis of the time asymmetry of knowledge is also compatible with our knowing very little about either the past or the future.

---

[7] Horwich argues that "the phenomenon of recording is an instance of the pattern of events that is known . . . as a 'normal fork'" (1987, p. 85). He spells this out in the following passage:

> A recording system, S, gets into each of its informative states, $S1$, $S2$, . . . , much more often than it gets into its noninformative states, those that are not associated with any particular environmental circumstances. And this heavy clustering constitutes a correlation that is explained by the frequent presence of prior circumstances $C1$, $C2$, . . . Thus, the association of S being in informative state Sk and prior condition Ck, which is essential to the performance of recording systems, is an instance of the general fact that correlations are causally explicable. (Horwich 1987, p. 85)

I find this to be rather obscure (and so does Savitt 1990), and Horwich gives no examples to help us see what it would mean for a recording device to get into each of its informative states "much more often" than it gets into its noninformative states. Fortunately, in the present context, nothing much rides on the question whether the asymmetry of recording and precording devices can be reduced to the fork asymmetry.

[8] Horwich makes a fascinating attempt to link his reflections on recording devices with the findings of physical cosmology. His view, roughly, is that the asymmetry of recording and precording systems can be explained in terms of the fork asymmetry, and that the fork asymmetry, in turn, can be explained by reference to an asymmetry having to do with "cosmic input noise" (Horwich 1987, pp. 71–76, 88–90). By this he means a randomness in the initial conditions of the universe. This cosmic input noise is also time-asymmetrical.

Second, the two proposed explanations of the time asymmetry of knowledge also help us to understand how natural science can deliver any knowledge of the past at all. It is only because of natural recording systems, and because correlations usually have common causes, that we can reasonably claim to know anything about the past.

1.3 THE PAST *VS.* THE MICROPHYSICAL

The time asymmetry of knowledge provides an excellent model for thinking about epistemic asymmetries in general. This book explores a different (more controversial, and less widely appreciated) epistemic asymmetry between the past and the microphysical: There is something that limits or obstructs our knowledge of prehistory, but not our knowledge of present microphysical entities, events, and mechanisms, at least not to the same extent. Just as we can know more about the past than about the future, we can know more about the tiny than about the past.

In distinguishing between the past and the microphysical, I aim to call attention to the different factors that can render entitites, processes, and events unobservable to us. Some things are unobservable because they existed or occurred long ago; other things are unobservable owing to their small size relative to us. Some things – e.g. the electrons of the dinosaurs – are unobservable on both counts. It is also worth noting that sometimes small size is not the only thing that makes microphysical entitites unobservable. The particles described by fundamental physics have properties, such as spin and polarization, that are just not the sorts of properties that human sense organs could detect. Those particles also lack determinate locations, and have other features that make them unobservable. For these reasons, we might wish to say that the unobservability of elementary particles is overdetermined. In chapter 3, I clarify these issues further, and provide more justification for the classification of unobservables as "past" or as "tiny." For now, let's just suppose, for the sake of argument, that the classification will hold up under scrutiny. Why should there be an epistemic asymmetry between the past and the tiny?

First, as Ian Hacking (1983) has emphasized, scientists can and do use experimental apparatus to manipulate tiny things and events. With the help of technology, it is possible to intervene in the microphysical world. Hacking argues that the best reason for thinking that electrons and positrons (for example) really do exist is that we can do things with them, and can even use them in the construction of tools for the detection of other unobservables, such as quarks with fractional electric charges:

> Moreover, it is not even that you use electrons to experiment on something else that makes it impossible to doubt electrons. Understanding some causal properties of electrons, you can build a very ingenious complex device that enables you to line up the electrons the way you want, in order to see what will happen to something else . . . Electrons are no longer ways of organizing our thoughts or saving phenomena in some other domain of nature. Electrons are tools. (Hacking 1983, p. 263)

Our ability to treat tiny things as tools is crucial to understanding how it is possible for us to have scientific knowledge of the microphysical world – just as the existence of recording devices is crucial to understanding how we can have scientific knowledge of the past. The experimental manipulation of microphysical entities and events makes it possible for scientists to test, and in some cases, confirm new theories. We cannot, however, manipulate things and events that existed and occurred long ago. This may seem like a trivial and uncontroversial point, just as it may seem obvious that we have an abundance of recording devices and a dearth of precording devices. However, this *asymmetry of manipulability* means that there is something – namely, our inability to intervene in the past – that limits our knowledge of the past without so limiting our knowledge of the tiny.

The second source of the epistemic asymmetry between the past and the tiny has to do with the different roles that background theories can play in science. In general, a background theory is a well-established theory that scientists take for granted when working on a problem in a related area. In some cases, background theories may serve to limit our scientific ambitions, because they give us reason to think that certain kinds of evidence will never become available. In other cases, though, background theories may serve to enlarge our scientific ambitions by showing us how to create new kinds of evidence. We might call these two possible roles for background theories the *dampening role* and the *enlarging role*, respectively. To give a couple of examples: Theories of optics have often played the enlarging role, because they have enabled scientists to devise ever more powerful microscopes and telescopes, thus expanding the range of observable evidence against which to test their theories. On the other hand, theories of taphonomy (which is the study of the fossilization process) have more often played the dampening role, because they imply that a great deal of evidence concerning past life on Earth has been destroyed forever. Theories about the past do not always play the limiting role. For example, background theories about the past often tell us what kinds of recording systems there are.[9] But even so, background theories about the past seldom, if ever, tell scientists how to create new empirical evidence, which is to say that they seldom, if ever, play the enlarging role. By contrast, background theories about the microphysical world frequently do tell us how to create new evidence by which to test claims and theories. We can call this the *role asymmetry of background theories*. It is closely related to the previous one, because often the way to create new evidence is by manipulating conditions in the laboratory.

My thesis, then, is that there is a rough sense in which we can know more about the tiny than about the past. All I mean by that is that certain factors (our inability to manipulate the past, as well as the dampening role played by historical background theories) limit our knowledge of the past but not our knowledge of the tiny. Or to put it another way: Our ability to manipulate tiny things and events helps us a great deal in our endeavors to acquire knowledge of the microphysical structure of the universe. But if we seek knowledge of the past, we will have to do without this help. These are the limits that Wilson and Carrano run up against in their work on titanosaurs.

This thesis needs to be qualified slightly. One could reasonably argue that at a certain point in their attempts to discern the microphysical structure of the universe, scientists run into much the same barriers that limit our knowledge of prehistory. Perhaps at some point, things get so unbelievably small that we can no longer manipulate them. And perhaps when we get down to a certain scale of smallness, background theories imply that we will probably never have the evidence that we would need to distinguish between rival hypotheses and theories. For example, many physicists worry about the testability of claims associated with string theory. Jarrett Leplin, a philosopher of science whose work I discuss at greater length in chapter 5, describes the current state of play in some areas of fundamental physics as follows:

> The new situation is that the very theories whose credibility is at issue themselves ordain their own nonconfirmability. If the latest theories are correct, then we should not expect to be able to confirm them. For they tell

---

[9] Peter Kosso (2001, pp. 61–64) makes this point nicely through his discussion of middle range theories in archaeology.

us, in conjunction with well-established background information, that the conditions under which the effects they predict occur are not technologically producible. (Leplin 1997, p. 178)

This is another way of saying that some theories of fundamental physics appear to play the dampening role. If Leplin is right, then there will be a point at which fundamental physics begins to look a lot like histori-cal science, for there is a point at which fundamental physics runs into the same limitations and obstructions. Beyond that point, fundamental physical theories (such as string theory) begin to look a lot more specula-tive. Many physicists today worry about how we could ever subject string theory to a risky empirical test.

So it is important to acknowledge that at some point, the scientific study of the microphysical world may run into limits that resemble the limits to our knowledge of prehistory. However, this is compatible with the existence of an epistemic asymmetry between the past and the tiny. The point that I will return to frequently in subsequent chapters is that *we cannot manipulate the past at all*, though we can manipulate lots of unobservably tiny things, from genes to electrons. In addition, few if any theories about the past ever play the enlarging role, but many theories about the tiny teach us new ways to create empirical evidence. Perhaps the best way to think about the asymmetry is to recognize that scientists can acquire a great deal of knowledge of the microphysical world without encountering these limits at all, even if they do eventually encounter them; historical researchers, by contrast, must deal with these limitations at every step of the way.

So far, I have introduced two "deep" asymmetries between the past and the microphysical world – the asymmetry of manipulability and the role asymmetry of background theories. I have also argued that these two deeper asymmetries give rise to the epistemic scope asymmetry between the past and the tiny. This closely parallels the situation described by Paul Horwich, in which the fork asymmetry and the asymmetry of recording and precording devices give rise to the time asymmetry of knowledge.

Why are these asymmetries so important? Why should scientists, philosophers, or anyone care so much about them? In order to appre-ciate what is at stake here, we will need to look carefully at the scientific realism debate. The main reason why the asymmetry of manipulability and the role asymmetry of background theories matter is that they have an impact on the ongoing discussion of scientific realism. I contend that the failure to take historical science seriously has caused that discussion to be skewed. Nor is this merely some minor oversight that needs to be corrected before we can move on to other things, because it has to do with what I see as a major disconnect between philosophers interested in realism and scientists working in fields such as geology, paleontology, evolutionary biology, and archaeology. Questions about scientific realism really do matter to our understanding of historical science, but since the main players in the realism debate seldom, if ever, have anything to say about prehistory, one could easily get the idea that scientific realism has nothing to do with historical science at all. I aim to show why scientists working on prehistory ought to care about the realism debate, just as much as I hope to show why philosophers working on the realism debate ought to care about historical science. I will do this by exploring the consequences of the asymmetries between the past and the microphysical.

1.4 SCIENTIFIC REALISM

What is scientific realism? First, some context: Philosophy has seen many "realism debates" over the years. During the middle ages, realist philosophers disputed with nominalists about the status of universals. Compare two red things – say, an apple and a fire hydrant. Is the redness that seems to exist in both of these particulars something real in its own right? Do we have three things here (the apple, the fire hydrant, and the redness) or just two? Nominalists held that there are only two things, the apple and the fire hydrant – namely, the two particulars. Scholastic realists held that the redness (i.e. the universal) exists, too. In the late 1600s and early 1700s, idealist philosophers such as Leibniz and Berkeley attacked John Locke's realist theory of perception. Locke held that we perceive only our ideas, and that those ideas represent things in the external world. According to this theory, my idea of the coffee mug on my desk represents the coffee mug, which exists "out there," independently of my mind. Berkeley argued on the contrary that objects like coffee mugs are ideas, and that their existence consists in their being perceived by a mind. Perhaps the earliest example of a realism debate in the western philosophical tradition occurs in Plato's dialogue, *Euthyphro*. There Socrates famously disagrees with Euthyphro concerning the nature of piety. Euthyphro, a relativist, suggests that what makes something pious is the fact that it is loved by the gods; Socrates, on the other hand, floats the realist hypothesis that what makes the gods love something is the fact that it is pious. Sometimes realism debates concern semantics. Take, for example, the sentence, "We have a duty to send aid to victims of natural disasters." Moral realists think that sentences such as this one have literal meaning, in the sense that they can be either true or false. Realists might also add that whether such a sentence is true or false depends on whether we do in fact have such a duty. But some philosophers have suggested that sentences such as this one may not have any literal meaning at all; instead, they serve only to express something – say, the speaker's feelings about natural disasters, or a pro-attitude toward sending aid. Some realism debates have more to do with metaphysics than with semantics, since they concern the existence of certain sorts of entities or properties. For example, many people think that there are such things as beliefs and desires. But some radical philosophers ("eliminative materialists") have argued that there are no such things.

This quick sampling shows that realism debates in philosophy have a tendency to cross over boundaries between metaphysics, epistemology, philosophy of language, and philosophy of mind. Therefore, in discussing any variety of realism, we will need to distinguish carefully among the different axes of potential disagreement between realist and nonrealist philosophers. It is also worth bearing in mind that philosophers have engaged in "realism debates" concerning just about everything imaginable: the self, the external world, God, numbers, universals, mental states, possible worlds, biological species, other people's minds, goodness, and so on and on.

Modern science has generated a new realism debate. One of the most central problems of the scientific realism debate is an *epistemological* problem, or a problem having to do with scientific knowledge: Virtually all scientists and philosophers of science are empiricists, in the sense that they think that all claims to scientific knowledge must be based on evidence, and that our evidence comes from observation and experimentation. Can observational evidence ever support our claims to have scientific knowledge of entities and mechanisms that we cannot, and probably never will observe? Does science deliver knowledge of unobservables?

It is important to stress that this epistemological problem is just one of several dimensions of the scientific realism debate, and I will discuss some of the other dimensions in later chapters. However, since in this book I am mainly interested in exploring the consequences that the two asymmetries between the past and the tiny might have for the realism debate, it makes sense to zero in on the epistemological dimension of that debate. For now, the important thing to see is that scientific realists unite in saying that we really can and do have scientific knowledge of unobservable entities and mechanisms. This epistemological optimism shows up again and again in recent formulations of the realist position:

> *Richard Boyd*: Scientific realists hold that the characteristic product of successful scientific research is knowledge of largely theory-independent phenomena, and that such knowledge is possible (indeed actual) even in those cases in which the relevant phenomena are not, in any non-question-begging sense, observable.[10] (Boyd 1990, p. 355)

Here Boyd combines epistemological optimism with a metaphysical claim about the theory-independence of the objects that scientists study.

> *Stathis Psillos*: The epistemic stance [of scientific realism] regards mature and predictively successful scientific theories as well-confirmed and approximately true of the world. So, the entities posited by them, or at any rate, entities very similar to those posited, do inhabit the world. (Psillos 1999, p. xix)

Here Psillos makes the following connection: If a theory says that there are quarks, and that quarks have such-and-such properties; if the theory is approximately true of the world; and if truth consists (as most realists think) in some sort of correspondence with the world, then those quarks (or entities very similar to them) do exist.

> *Ernan McMullin*: The basic claim made by scientific realism, once again, is that the long-term success of a scientific theory gives reason to believe that something like the entities and structure postulated by the theory actually exists. (McMullin 1984, p. 26)

This is another way of saying that the observable evidence can "give us reason to believe" certain claims about unobservables – or that in cases where a scientific theory enjoys long-term empirical success, we may reasonably take ourselves to know something about the entities and structure posited by that theory.

> *Jarrett Leplin*: [T]here are possible empirical conditions that would warrant attributing some measure of truth to theories – not merely to their observable consequences, but to theories themselves. This is minimal epistemic realism (MER), a relatively weak form of scientific realism that need not endorse any actual theory. (Leplin 1997, p. 102)

---

[10] The notion of unobservable phenomena is an oxymoron. Where Boyd uses the term "phenomena", we should take him to mean "things and events."

Leplin here affirms the possibility of scientific knowledge of unobservables, while leaving open for the time being the question whether we actually do have such knowledge in any particular case. This is the weakest possible version of realist epistemology of science.

Even scientific realists who seem to de-emphasize epistemology join this chorus. Thus, Michael Devitt, who argues that realism is most fundamentally a metaphysical view, characterizes the realist position as follows:

> *Michael Devitt*: Tokens of most current unobservable scientific physical types objectively exist independently of the mental. (Devitt 1991, p. 24)

Although Devitt's formulation of the scientific realist view is carefully designed to stress the metaphysical dimensions of that view, Devitt is also an epistemological optimist. The "unobservable scientific physical types" to which he refers include electromagnetic waves, electrons, protons, DNA molecules, and so on. Devitt clearly thinks that we can confidently claim to know that these things, rather than some entirely different things, exist.

The point of this brief survey is to show that although realists disagree amongst themselves about how best to characterize the scientific realist position, all realists share an epistemological optimism when it comes to our knowledge of unobservable entities and mechanisms. Some realists are more cautious, some more ambitious, but all share this optimism. All believe that there could be situations in which it is reasonable for us to take scientific claims about unobservables to be true or nearly true; most would go further to say that many current scientific claims about unobservables are true or nearly true.

On the other hand, it is important to note that not all who share this epistemological optimism are realists in the fullest sense. Some philosophers share the optimism while denying some of the other things that scientific realists typically want to say. For example, social constructivists usually agree with realists that we can have knowledge of unobservables; however, they reject the realists' claim that the world is what it is independently of us, our theories, and our conceptual schemes. Although few professional philosophers of science these days sympathize much with social constructivism, many scientific realists see themselves as reacting against the social constructivism that pervades much work in the history and social studies of science.[11] One other interesting example of a

---

[11] For helpful discussions and evaluations of social constructivism, see Hacking (1999); Ruse (1999); Kukla (2000); and Parsons (2001).

non-realist philosopher who shares the realists' epistemological optimism is Arthur Fine (1984). Even though Fine objects to just about everything else that scientific realists want to say, he does seem to think that we can have scientific knowledge of unobservables:

> I certainly trust the evidence of my senses, on the whole, with regard to the existence and features of everyday objects. And I have similar confidence in the system of "check, double-check, triple-check" of scientific investigation, as well as other safeguards built into the institutions of science. So, if the scientists tell me that there really are molecules, and atoms, and ψ/J particles, and, who knows, maybe even quarks, then so be it. I trust them, and thus must accept that there really are such things with their attendant properties and relations. (Fine 1984, p. 95)

Although Fine famously rejects realism in favor of a deflationary stance that he calls the *natural ontological attitude* (or NOA), passages like this one have led other philosophers (most notably, Musgrave 1989) to complain that Fine is a closet realist. These critics are partly right, partly wrong. Fine does share the realists' epistemic optimism, but he remains agnostic with respect to a range of other questions that realists have views about. For example, Fine thinks that both realists and social constructivists push inflated metaphysical pictures of science, with realists claiming that the world is independent of us, and constructivists claiming that it is somehow dependent on us, our theories, or our conceptual schemes. He argues instead that we should simply suspend judgment on this issue. Since Fine's work provides the inspiration for the natural historical attitude, I will have much more to say about his ideas in later chapters.[12]

So why is there any epistemological debate at all? Does anyone seriously doubt our ability to have scientific knowledge of unobservables? Over the last couple of decades, critics of scientific realism have employed a pair of skeptical arguments – the pessimistic induction from the history of science and the underdetermination argument – to challenge the realists' optimism concerning our knowledge of unobservables. Both of these arguments are intuitively easy to grasp. The first one begins with the recognition that on many past occasions, scientists took themselves to have knowledge of unobservables but later turned out to be badly mistaken. So there is a good chance that at some future time we will discover that our best current theories are also mistaken in what they say about the unobservable world. According to the second line of argument, there are

---

[12] For a helpful interpretive discussion of Fine's work, see Rouse (1996, ch. 2–3). Rouse emphasizes the postmodern and contextualist aspects of Fine's work.

several ways the unobservable world could be, given all the observable evidence. Our decisions concerning what to believe about the unobservable world are therefore underdetermined by the observable evidence. In the face of these powerful skeptical arguments, scientific realists clearly need to justify their epistemological optimism. Realists, for their part, have employed a variety of abductive arguments (or arguments to the best explanation) in order to defend their optimistic outlook. The most popular version of this strategy is the classical inference to the best explanation of the success of science: Our current scientific theories enjoy tremendous predictive success. This success would be a mystery if those theories were not true or nearly true. Much of the debate about scientific realism has focused on these arguments for and against epistemological realism.

So far, I have argued that all scientific realists are optimistic when it comes to scientific knowledge of unobservable entities and mechanisms, but not all who share this optimism are scientific realists in the fullest sense. Some thinkers who reject other parts of the realist picture (for example, social constructivists and NOAers such as Arthur Fine) share the realists' optimism.

Before going any further, however, we need to draw a sharper distinction between two strengths of epistemic optimism. On the one hand, there is the minimal optimism embodied in the claim that it is possible for us to have scientific knowledge of unobservables. Yet it is one thing to ask whether we can know anything at all about things that we cannot observe – optimists will say "Yes!" It is another thing to ask how much we already do know about things that we cannot observe – optimists will say "Quite a lot!" This second kind of optimism about the scope of scientific knowledge of unobservables is stronger than the more basic optimism about the possibility of such knowledge. Most scientific realists exude both kinds of optimism. Some even build optimism about the scope of knowledge into their formulations of the realist position. For example, when Michael Devitt writes that "Tokens of most current unobservable scientific physical types objectively exist independently of the mental," he is, I think, asserting that we know quite a lot about which physical things really exist. On the other hand, some realists, such as Jarrett Leplin, take care not to include any claims at all about the scope of knowledge in their formulations of the realist position. Leplin defines "minimal epistemic realism" as the view that "[T]here are possible empirical conditions that would warrant attributing some measure of truth to theories – not merely to their observable consequences, but to theories themselves." Notice that this says nothing at all about how often those empirical conditions are satisfied. From this point onward, when I refer to "epistemic realism," I will try to make clear whether I am talking about optimism with respect to the possibility, or with respect to the scope of scientific knowledge.

Why is this distinction between questions about the possibility and questions about the scope of scientific knowledge so critical? Remember that my project is to trace out the logical consequences of the asymmetry of manipulability and the role asymmetry of background theories. Some of those consequences have to do with the scope of knowledge. Indeed, the epistemic asymmetry between the past and the tiny is an example of an epistemic scope asymmetry. But I will argue that the asymmetry of manipulability and the role asymmetry of background theories also make a difference to the arguments that philosophers use to try to establish the possibility of scientific knowledge of unobservables (See diagram).

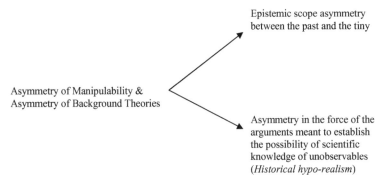

I will argue that the asymmetry of manipulability and the role asymmetry of background theories actually give rise to two different epistemic asymmetries – one concerning the scope of knowledge, and one concerning the possibility of knowledge. Most realists rely on some form of inference to the best explanation to establish the possibility of scientific knowledge of unobservables. In chapter 3, I argue that the asymmetry of manipulability and the role asymmetry of background theories mean that these standard arguments for realism (if they are any good at all) have less force with respect to prehistory than they do with respect to microphysics. We can call this further consequence of the two asymmetries *historical hypo-realism*. The term is meant to suggest that the arguments for thinking we can have knowledge of the past are less good than those for thinking we can have knowledge of the microphysical.

Historical hypo-realism is a comparative claim to the effect that the strength of certain realist arguments varies from one context to another, and as such it is compatible with the view that we should be minimal epistemic optimists about the past as well as about the tiny. Saying that the abductive case for realism about the past is weaker than that for realism about the tiny is compatible with saying that the arguments for both views are very good. Historical hypo-realism is simply a claim about the relative strength of certain arguments that realist philosophers like to make.

To sum up the important claims of this section: All scientific realists (and a few others, like Arthur Fine), think that scientific knowledge of unobservables is possible. Most also take an optimistic view of the scope of our actual knowledge of unobservables. The burden of chapters 2 through 5 is to show that the asymmetry of manipulability and the role asymmetry of background theories mean that both of these forms of optimism about scientific knowledge need to be qualified. Thus, the argument of this book has two prongs. Chapter 3 will focus on the possibility of knowledge of unobservables, while chapters 2, 4, and 5 will focus on questions about the scope of scientific knowledge.

1.5 A SKEWED DEBATE

Imagine a fictional universe containing a planet somewhat like ours. On that planet live a number of investigators. The investigators are able to use their sense organs, together with technological aids, to observe some of the things in their universe, but many more things lie beyond their powers of observation. Moreover, in this fictional world, there are two basic kinds of things that are unobservable to the investigators – things of kind K, and things of kind K^*. (For the moment, do not worry about what these two kinds really are.) The investigators on this planet divide their labor: some of them study things of kind K, and others study things of kind K^*.

On this fictional planet, there are also some philosophers who take an optimistic attitude toward the activities of the investigators. For complicated historical reasons, the philosophers spend most of their time thinking about the achievements of the investigators who focus on things of kind K. These philosophers, who call themselves "realists," claim that the investigators have succeeded in acquiring knowledge of things that no one can observe. (Some other philosophers, who disagree with the realists, deny this claim. For now, though, do not worry about who those other philosophers are, or why they disagree with the realists.) The realists in this fictional world exhibit a general epistemic optimism about the investigators' ability to deliver knowledge of things that no one can observe. They think both that it is possible for the investigators to have knowledge of unobservables, and that the investigators have already acquired a good deal of such knowledge.

Next, suppose for the sake of argument that the differences between objects of kind K and objects of kind K^* *make an epistemic difference*. Maybe things of kind K^* have distinctive features that make them considerably more difficult to study. Perhaps these features create obstacles to the investigation of things of kind K^* that are not present with respect to things of kind K. These relevant differences between K and K^* imply that one ought to have more epistemic optimism (in both of the above senses) with respect to things of kind K, and less optimism with respect to things of kind K^*. However, the realist philosophers on this fictional planet do not express differential optimism with respect to K and K^*, because they have not noticed that the differences between K and K^* make an epistemic difference. The critics of realism, who doubt that the investigators can deliver any knowledge at all that goes beyond the limits of what they can observe, also overlook the epistemically relevant differences between K and K^*. In this story, the realist philosophers and their opponents spend much of their time exchanging generic arguments about the possibility and scope of knowledge of unobservables.

There are two ways in which the philosophical debate in this story can become skewed. The first involves a genus/species confusion. Suppose, for example, that one particularly ambitious realist philosopher makes an argument to establish the possibility of knowledge of entities of kind K, and then concludes that he has thereby established the possibility of knowledge of unobservables. Because there are, let us suppose, epistemically relevant differences between the two species of unobservables K and K^*, conclusions about K do not necessarily translate into conclusions about unobservables. I will examine this type of genus/species confusion with greater care in section 3.3. For now, the important thing to see is that since the philosophers in our story are most familiar with the achievements of the investigators working on K, it will be awfully easy for them to conflate claims about things of kind K (i.e. about the species) with claims about unobservables (i.e. about the genus). Clearly, the way to avoid mistakes of this sort is simply to call attention to the relevant differences between K and K^*.

The second way in which the debate in this story gets skewed does not involve any fallacious inferences. The problem is just that the philosophers in this story carry out their discussion at too high a level of generality. That is, the debate is skewed in the sense that it takes place at the genus level when it should take place at the species level. Perhaps the best way to bring this out is with the help of a simple analogy. When the philosophers in this story ask, "Is it possible for the investigators to acquire knowledge of unobservables? And if so, how much knowledge have they acquired?" that is a bit like asking, "Is it possible for mammals to digest meat? And if so, how much meat do mammals in fact eat?" It's not that these are necessarily bad questions – we can certainly pose them if we want to. But the questions do obscure species level differences (e.g., dietary and physiological differences among species of mammals) that turn out to be highly relevant to what is being asked about. In this example, if the biologists never bothered to look at relevant differences among species of mammals, we should say that their answers to the more generic questions about mammals are seriously incomplete. The same goes for the philosophers in this story: They can try to answer the generic questions about knowledge of unobservables, but if they never look at the relevant differences between K and K^*, they will miss out on something important.

You may have guessed by now that in my view, the "fictional" situation I have just described is not really fictional at all. I have just described what I see as the current state of play in the scientific realism debate. The two kinds of unobservables, K and K^*, are the past and the microphysical. The differences between them that make an epistemic difference are the asymmetry of manipulability and the role asymmetry of background theories.

2

The colors of the dinosaurs

In my study I have a black-and-white photograph of my grandfather as a young man, standing in front of a house holding a lunchbox. I sometimes wonder what, if anything, was in the lunchbox. That is a simple question about the past that no one will ever be able to answer. Many questions in historical science are like that: for instance, asking about the colors of the dinosaurs is just like asking what was in my grandfather's lunchbox. In this chapter, I argue that these unanswerable questions – which I will call *local underdetermination problems* – are more common in historical science than in experimental science. This is one consequence of the role asymmetry of background theories.

In chapter 1, I described one example of an asymmetry in time: the time asymmetry of knowledge. In this chapter, I will begin with another alleged asymmetry in time: the asymmetry of overdetermination. Carol Cleland (2002) has recently invoked David Lewis's (1979) thesis of the time asymmetry of overdetermination in order to answer the charge that prototypical historical science is epistemically inferior to classical experimental science. Cleland argues that the asymmetry of overdetermination is a fact about our universe that underwrites the distinctive methodologies of historical and experimental science, guaranteeing that the one methodology is, epistemically speaking, just as good as the other. In this chapter, I argue that Lewis's notion of the asymmetry of overdetermination cannot do the work that Cleland wants it to do. I then go on to reinforce some of the claims made in chapter 1 by showing that historical science is, in at least one interesting sense, epistemically disadvantaged relative to experimental science.[1]

---

[1] For a somewhat different critique of Cleland, see Kleinhans, Buskes, and de Regt (2005). They also take the view that underdetermination is especially common in earth science.

2.1 LEWIS ON THE ASYMMETRY OF OVERDETERMINATION

David Lewis defines the "determinant" of any fact about the world as "a minimal set of conditions, jointly sufficient, given the laws of nature, for the fact in question (members of such a set may be causes of the fact, or traces of it, or neither)" (1979, p. 474). A fact or affair is overdetermined just in case it has more than one determinant at a given time. Overdetermination, Lewis suggests, is a matter of degree. A fact may have two or three determinants, or many more.

There are some familiar examples of earlier affairs overdetermining later affairs. For instance, when a convict is shot by firing squad, the death is overdetermined. Only one shot would have been sufficient to kill him. For another example, suppose that Jones shakes hands with three different people, all of whom have the flu. Jones's getting sick is overdetermined by these three handshakes. Although such cases show that earlier affairs sometimes overdetermine later affairs, Lewis thinks that cases like this are uncommon. Moreover, in most of these cases, the number of determinants is quite small. On the other hand, Lewis argues that overdetermination of earlier affairs by later affairs is both more common and more extreme: *"We may reasonably expect overdetermination toward the past on an altogether different scale from the occasional case of mild overdetermination toward the future"* (Lewis 1979, p. 474, my emphasis). Call this highlighted claim the "thesis of the time asymmetry of overdetermination." Notice that this is strictly a metaphysical thesis, a thesis about the nature of time and reality. There are a couple of different questions we might ask about this thesis. First, is it true? Second, does it have any interesting epistemological consequences?

Is overdetermination really asymmetrical, as Lewis suggests? Lewis's thesis is initially plausible. Suppose, for example, that two different people throw baseballs at the same window at the same time. In that case, the shattering of the window is overdetermined by earlier events, but such cases are relatively rare. Cleland points out that the breaking of the window is overdetermined by numerous subcollections of shards of glass lying on the kitchen floor (2002, p. 487).[2] That overdetermination of earlier facts by later traces occurs whenever a window breaks. Beyond

---

[2] This is closely related to the disparate trace hypothesis, which I defend in chapter 8. In fact, the disparate trace hypothesis may just be a restatement of one half of the asymmetry of overdetermination – i.e. of the claim that earlier events are usually overdetermined by later events.

pointing to examples like this, it is not clear to me how one would go about defending (or, for that matter, criticizing) such a thesis. This is especially true, because Lewis's thesis admits of exceptions: he only claims that earlier affairs *seldom* overdetermine later affairs, and that later affairs *usually* overdetermine earlier affairs.

For purposes of this chapter, I will assume that Lewis's thesis of the asymmetry of overdetermination is true. I shall argue, however, that Cleland is wrong to think that this metaphysical thesis has any interesting epistemological consequences. More specifically, she is wrong to suppose that the asymmetry of overdetermination "underwrites" the distinctive methods of prototypical historical science and classical experimental science, in the sense of guaranteeing that neither methodology is epistemically better than the other. Lewis, I think, hints that this thesis has no epistemological consequences when he says that "Most of these traces are so minute or so dispersed or so complicated that no human detective could ever read them" (1979, p. 474).

2.2 CLELAND'S ARGUMENT

According to Cleland, the methods of prototypical historical science differ from those of classical experimental science. Historical scientists proceed in roughly the following way:

1. Observe and describe puzzling traces of long-past events.
2. Postulate a common cause of those traces. The common cause is usually some token event or process that occurred long ago.
3. Test this hypothesis about the distant past against rival hypotheses by searching for a "smoking gun," or a present trace that, together with the other traces observed so far, is better explained by one of the rival hypotheses than by the other. (Cleland 2002, p. 481)[3]

As explanatory hypotheses proliferate, scientists search for smoking guns that will discriminate among them. Cleland gives a number of convincing examples of smoking guns in historical science. For instance, the presence of iridium and shocked quartz at the Cretaceous–Tertiary boundary, not

---

[3] Although Cleland does not discuss the notion of predictive novelty, it is interesting to compare her notion of a smoking gun with that of a novel prediction, which I discuss in chapter 5. The notion of a smoking gun seems to be the weaker of the two. It is what you get when you take Leplin's (1997) uniqueness condition all by itself, without his independence condition.

to mention the Chicxulub crater in Central America, are smoking guns for the Alvarez hypothesis that an asteroid impact triggered the extinction of the dinosaurs (Alvarez et al. 1980). None of the other potential explanations of the Cretaceous–Tertiary mass extinction imply the existence of a crater. Another example of a smoking gun might be the recent discovery of two shelled eggs inside a dinosaur, *Sinosauropteryx*, which confirms the hypothesis that some dinosaurs had paired oviducts (Sato et al. 2005).

It is worth pausing to make explicit one assumption that Cleland seems to make concerning the nature of scientific confirmation. Some philosophers of science hold that confirmation is always comparative. According to this view, it does not make sense to ask how well the evidence *E* supports some hypothesis *H*, except in reference to one or more hypotheses in competition with *H*. Thus, a comparativist would insist that the discovery of the Chicxulub crater counts in favor of the Alvarez asteroid impact hypothesis only by comparison with number of other potential explanations of the K–T extinction, such as the hypothesis that the dinosaurs were wiped out by an epidemic. The evidence does not support the Alvarez hypothesis, considered in isolation from these competitors. For ease of exposition, I will follow Cleland in working from the assumption that confirmation is always comparative. None of the arguments of this chapter depend on this assumption, however.[4]

Cleland shows that historical scientists exploit the asymmetry of overdetermination in the following way: The thesis of the asymmetry of overdetermination implies that most events in the past will have a large number of determinants at the present time, where each determinant is a set of conditions (or traces) that, together with the laws of nature, are jointly sufficient for the earlier event. This explains how the distinctive methodology of historical science can deliver scientific knowledge of the past.

Next, Cleland contrasts prototypical historical science with classical experimental science. She emphasizes that since these are ideal types, a particular piece of scientific work may be partly historical, partly experimental. Historians often reason experimentally, and experimentalists sometimes reason historically. According to her, practitioners of classical experimental science (an ideal type) proceed in the following way:

---

[4] One anonymous reviewer wondered whether the differences between historical and experimental science have any impact on the plausibility of this comparativist view of confirmation. I have not been able to think of any reason why comparativism should be any more or less plausible in historical than in experimental science.

1. Begin by forming a hypothesis about a regularity among event types.
2. Predict what will happen if the hypothesis is true, and if a given test condition is realized.
3. Run a series of experiments in which conditions are manipulated so as to rule out false positives and false negatives.

For example, suppose that an ecologist wants to test a hypothesis about the effects of deer browsing on local vegetation. She makes a prediction about what sorts of plants would grow in a given spot, were they not browsed by deer, and she tests this hypothesis by fencing off a small plot of forest and waiting to see what happens. The ecologist then repeats the experiment while varying certain conditions, such as the amount of sunlight available to the plants or the acidity of the soil, by fencing off different plots in different places – for example, one on top of a dry ridgeline and another in a shady ravine.

Cleland argues that the experimental method is an attempt to cope with or even circumvent the time asymmetry of overdetermination. In order to make this point about experimental science, she relies on the following example: A short circuit is not sufficient for the occurrence of a destructive fire. It is only a partial cause. In order for the fire to occur, there must be flammable materials nearby, the sprinkler system must malfunction, and so on. The burning down of the house is therefore causally *under*determined by the short circuit.

Suppose that after the ecologist fences off a certain patch of woods to prevent deer browsing, saplings of a given tree species begin to flourish in the protected area. The ecologist still needs to consider other possible causal influences. What if some other animal, aside from the deer, has been destroying the saplings? And what if the fence also succeeds in keeping out that other animal? In that case, the experimental results might yield a false positive. Thus, "there is a need to ferret out and control for additional factors that are relevant to the total causal situation," and that is just what experimentalists do when they manipulate test conditions (Cleland 2002, p. 494).

Cleland's argument, then, can be summarized as follows:

P1. Later affairs usually overdetermine earlier affairs, but earlier affairs usually underdetermine later affairs.
P2. Historical scientists exploit one half of this asymmetry: their methods for testing hypotheses about past event tokens are appropriate because later affairs usually overdetermine earlier affairs.

41

P3.   The experimental method is a strategy for coping with the other half of this asymmetry: Since earlier affairs and events (such as the short circuit) usually underdetermine later affairs and events (such as the burning down of the house), anyone who wishes to test hypotheses about regularities among event types must run a series of trials in which different test conditions are manipulated, with the aim of ruling out false positives and false negatives.

C.   Therefore, prototypical historical science and classical experimental science are equally good, epistemically speaking.

This is an ingenious argument, and Cleland does an excellent job making the case for P2 and P3. I think she is probably right that historical and experimental science exploit different aspects of the time asymmetry of overdetermination, and by pointing this out, she has contributed a great deal to our understanding of the relationship between the two methodologies. My goal in this chapter is to explain why the epistemological conclusion does not follow from the premises.

At certain points in her paper, Cleland shifts from talking about overdetermination as defined by Lewis to talking about epistemic overdetermination. She says, for example, that "the asymmetry of (epistemic) overdetermination is ultimately founded on a time asymmetry of nature" (2002, p. 489). She also says that "the overdetermination of causes by their effects is (strictly speaking) only epistemic" (2002, p. 488). As we have seen, Lewis's thesis about the asymmetry of overdetermination is a metaphysical one. For Lewis, the determinant of any affair can be either a set of earlier causes, or a set of later traces, and the determination relation is a relation among affairs (or facts), not a relation between hypothesis and evidence. Perhaps Cleland is misled into thinking that the conclusion follows from the premises stated above because she fails to distinguish clearly between causal/metaphysical overdetermination of the sort that Lewis is talking about and epistemic overdetermination.

What is epistemic overdetermination? One initially plausible suggestion is that a hypothesis or theory *H* is epistemically overdetermined just in case there are at least two distinct arguments, or lines of evidence, each of which alone is sufficient to justify believing *H*. For example, someone (not me) might think that belief in the existence of God is epistemically overdetermined by the various arguments for God's existence, because each of those arguments, by itself, would be sufficient to justify belief in God. I am not sure if this is what Cleland means by epistemic overdetermination.

At any rate, I will argue that the asymmetry of overdetermination does not imply an asymmetry of epistemic overdetermination, in this sense of epistemic overdetermination. On the contrary, metaphysical overdetermination, in Lewis's sense, is compatible with epistemic underdetermination.

At one point in her paper, Cleland says that the asymmetry of (causal? epistemic?) overdetermination could be probabilistic:

> Although Lewis characterizes the asymmetry of overdetermination in terms of sufficiency, it could turn out to be a probabilistic affair, with the ostensibly overdetermining subcollections of traces lending strong but, nevertheless, inconclusive support for the occurrence of their cause. Like the determinism in Lewis's original version, the probabilistic support offered by collections of traces for hypotheses would be an objective feature of the world. (Cleland 2002, 490)

Elsewhere she refers to this probabilistic phenomenon as "the asymmetry of (quasi) overdetermination" (Cleland 2002, 491). I have four distinct worries about this passage. First, it is hard to tell whether this probabilistic overdetermination is an epistemic or a causal notion. The claim that it is "an objective feature of the world" suggests a causal/metaphysical notion, as in Lewis's original version. However, the reference to "probabilistic support" suggests an epistemic notion. Second, how is quasi-overdetermination different from quasi-underdetermination? Why use the word "overdetermination" at all where we are not talking about sufficiency, as in Lewis's original version? Third, what reason is there to think that this probabilistic quasi-overdetermination is asymmetrical? Perhaps earlier affairs quasi-overdetermine later affairs to the same extent that later affairs quasi-overdetermine earlier ones. It is possible that quasi-overdetermination is asymmetrical, but further argument is needed to support this new thesis. Finally, there are still plenty of non-trivial cases of local epistemic underdetermination in which later traces do not even lend probabilistic support to hypotheses about earlier events.

In order to see why causal/metaphysical overdetermination does not imply epistemic overdetermination, and why it is compatible with epistemic underdetermination, we need only look at a case in which there is both metaphysical overdetermination and epistemic underdetermination. I will use one of Cleland's own examples.

43

2.3 WHY CAUSAL/METAPHYSICAL OVERDETERMINATION DOES NOT RULE OUT EPISTEMIC UNDERDETERMINATION

Cleland uses the example of a baseball shattering a window in order to illustrate Lewis's thesis of the asymmetry of overdetermination. The baseball hitting the window does not overdetermine the later traces (i.e. the shards of glass landing on the kitchen floor), but there are many subcollections of traces that overdetermine the baseball's hitting the glass, in Lewis's sense of overdetermination.

Notice, though, that even in this example, metaphysical overdetermination is perfectly compatible with epistemic underdetermination. Suppose we ask how many miles per hour the basball was traveling when it hit the window. Rival hypotheses about the baseball's speed, its angle of impact, the distance traveled, whether it was thrown or hit with a bat, etc., may all be underdetermined by the shards of glass lying on the floor.

Now suppose we develop the thought experiment a bit further. The owners of the house sweep up the shards, toss the baseball in the closet, and eventually repair the window. A few weeks later, the only traces of the event that remain are a few shards of glass underneath the refrigerator. The housecleaning and repair are examples of what Sober (1988, p. 3) calls *information destroying processes*. Consider the epistemic situation of the historical investigator who finds the shards of glass under the refrigerator. The investigator may grasp that they are traces of some sort, without having any idea what they are traces of. Are the shards the remains of a broken window, a broken wine glass, or a broken picture frame? Even if the historical investigator recognizes the traces for what they are, rival hypotheses about earlier events and processes will often be underdetermined by the available traces. After studying the shards under the refrigerator, the historical investigator will be completely stymied: The evidence does not permit her to discriminate at all between incompatible rival hypotheses (window *vs.* wine glass, football *vs.* baseball, etc.). Moreover, since the investigator knows that people usually clean up the mess when things like windows and wine glasses break, she has good reason to think that she will never find any traces that will enable her to distinguish between the rival hypotheses. In other words, she confronts a local epistemic *under*determination problem.

Or does she? Cleland might point out that the processes of clean-up and repair will leave traces of their own – a receipt for the window filed away somewhere, tiny pieces of glass stuck in the bristles of the broom, and so on. This is true, but unhelpful. Suppose the historical investigator finds some small bit of evidence suggesting that a window was shattered. Did a football or a baseball do the damage? Instead of allowing the researcher to investigate the scene a few weeks after the fact, make the investigator wait for a few decades, until all the traces of the clean-up have been cleaned up, scattered, or destroyed. What this shows, I think, is that Lewis's thesis of the asymmetry of overdetermination does not rule out epistemic underdetermination. This is precisely the sort of case in which the "traces are so minute or so dispersed or so complicated that no human detective could ever read them" (Lewis 1979, p. 474). Lewis's thesis of the asymmetry of overdetermination is compatible with the epistemological thesis that local underdetermination problems are widespread in historical science. Indeed, I will argue that there is reason to think that local epistemic underdetermination is a bigger problem in historical than in experimental science, and so there is reason to think that Cleland's conclusion (C, above) is false.

Cleland's example of the Chicxulub crater, which is a smoking gun for the hypothesis that an asteroid collided with the earth approximately 65 million years ago, is typical of historical science in one way, but not in another. It is typical of historical science as far as methodology goes, because the scientists in this case sought to test their hypothesis by finding a smoking gun, just as Cleland describes. But it is atypical of historical science as far as epistemology goes. The event in question was of such a magnitude, and happened so recently (65 million years is not so long ago, geologically speaking) that its presently observable traces are a dead giveaway, just as the shards of glass and the baseball on the floor would be a dead giveaway to any investigator who happened on the scene before the homeowners had repaired the window and cleaned up the mess. It would be a mistake to infer from this sort of example that earlier causes are usually, or even very often epistemically overdetermined by their effects.

One potential objection at this point is that the example of the baseball shattering the window is misleading because it involves human agency. One might reasonably think that in nature, there is no one to "clean up after" geological events, and nothing analogous to the person who repairs the broken window. Why use a hypothetical scenario involving human agency when we are mainly interested in prehistory? I have several responses to this worry: First, the example of the ball shattering the window does show that overdetermination of earlier affairs by later affairs (in Lewis's sense) is compatible with epistemic underdetermination, which is all that I have aimed to show so far. Second, the example is Cleland's own. I hope to have shown that even in the case that she herself uses to illustrate the time asymmetry of overdetermination, earlier events can be (epistemically) underdetermined by their later traces. Third, information destroying processes in nature erase historical traces just as clean-up and repair erase the traces of the collision of the baseball with the window. Whether or not the traces are destroyed as a result of human agency is inessential to the argument. What matters is that our background theories give us reason to believe that they have been destroyed.

2.4 LOCAL UNDERDETERMINATION PROBLEMS IN
HISTORICAL SCIENCE

Historical scientists frequently find themselves in situations similar to that of the investigator who discovers a few shards of glass under the refrigerator. In order to show this, I will begin by offering an analysis of local epistemic underdetermination problems; then I will describe four cases from historical science that fit the analysis.

We can start with the notion of empirical equivalence: Two incompatible theories or hypotheses, H and H^*, are empirically equivalent just in case they have the same empirical consequence class, which is to say that they have exactly the same consequences with respect to the observations that scientists could ever make.

There are plenty of trivial cases of empirical equivalence in historical science. For example, what color were the dinosaurs? On p. 138 of David Norman's popular book, *The prehistoric world of the dinosaur* (just one example among many), there is a picture of a gray pachycephalosaur with a neon blue patch on the top of its head. At the beginning of the book, Norman writes that it is "difficult – in fact, almost impossible – to know what colors dinosaurs were" (1988, p. 8). Why only almost? Norman adds that it is possible to make guesses based on analogies with living organisms. (But see chapter 4 for discussion of some of the pitfalls.) For example, most big herbivores – elephants, rhinoceroses, and hippopotamuses – have dull grayish colors. Perhaps the same was true of the big herbivores of the Mesozoic. Nevertheless, anyone can see that the hypothesis that *Pachycephalosaurus* had a neon blue patch on its head is empirically equivalent to the hypothesis that it had a neon green patch. We have good reason to think this because we know that information about coloration is destroyed by the fossilization process. Our background theories of taphonomy tell us that we will never find any historical traces that render either of these hypotheses more probable than the other.

Can we be sure that the rival hypotheses about the color of *Pachy-cephalosaurus* are empirically equivalent? Suppose that in the future we encounter an extraterrestrial civilization that sent a zoological expedition to Earth many millions of years ago to conduct a detailed survey of the earth's flora and fauna, and that these extraterrestrials possess color photographs of pachycephalosaurs. One might reasonably argue that since we cannot rule out this possibility, we cannot be sure that the rival hypotheses about the color of *Pachycephalosaurus* are empirically equivalent. However, we still have no reason at all to think that we ever will come to possess such photographs. Furthermore, our background theories about taphonomy do give us good reason for thinking that all the information about the colors of dinosaurs has been completely destroyed, and therefore considerable justification for thinking that rival hypotheses about the color of *Pachycephalosaurus* are empirically equivalent. In this case, although we cannot be certain that H and H^* are empirically equivalent, because we cannot rule out the possibility of an encounter with alien dinosaurologists, our background theories nevertheless give us good reason for thinking that they are.

A *local underdetermination problem* is any situation in which the following conditions are satisfied:

a.   Two incompatible hypotheses, H and H^* are genuine rivals.
b.   As best anyone can tell, H and H^* have roughly equal portions of non-empirical theoretical virtue (simplicity, explanatory power, and the like).
c.   Background theories give us some reason to think that H and H^* are also empirically equivalent – i.e. that they have exactly the same observational consequences.

When these conditions are met, scientists ought to suspend judgment with regard to H and H^*. This could mean that they continue to search for a smoking gun that will discriminate between the two, even if there is no good reason to think that such a smoking gun will ever turn up. Or it could mean that they simply move on to more tractable research questions. It is easy to see how philosophers of science could underestimate the pervasiveness of local underdetermination problems in historical science – as I think Cleland does – because scientists themselves tend not to dwell on such problems. For this reason also, examples of local underdetermination problems in historical science are likely to seem a little contrived. No serious scientist would spend time looking for a smoking gun to distinguish between rival hypotheses about the colors of the dinosaurs, because there is good reason to doubt the existence of any such clues. Indeed, historical scientists are trained to identify local underdetermination problems and to move on to more tractable research questions. Thus, it would be difficult to produce examples of research problems that scientists are currently working on, and that clearly satisfy the above conditions for a local underdetermination problem.

Much of the discussion of underdetermination has focused on what might be called the *global underdetermination problem*.[5] The global problem is generated by the empirical equivalence thesis that for any hypothesis *H*, there is at least one empirically equivalent rival. Some philosophers have even suggested that for any hypothesis *H*, there are indefinitely many empirically equivalent rivals. In fact, it is easy to show that the following is true:

> *Strong Historical Empirical Equivalence Thesis.* For any hypothesis about the past *H*, there are indefinitely many empirically equivalent rivals.

We can generate the rivals algorithmically, in the following way: Consider the hypothesis that God created the universe at some past time *t* (six seconds ago; six minutes ago; six thousand years ago; six trillion years ago, etc.), and that when he did so, he made the universe to appear older/younger than it really is.[6] Since there are indefinitely many past times that we can plug in for *t*, we can form indefinitely many creationist hypotheses, each of which will be empirically equivalent to any other historical hypothesis we care to dream up. (See also the algorithms proposed by Kukla 1996.)

Many philosophers of science dismiss these radical skeptical worries on the grounds that the algorithmically generated theories are not genuine rivals of the original scientific theories.[7] Whereas local underdetermination problems arise during the course of scientific inquiry, global underdetermination problems seem to be imposed upon science by philosophers. Anyone who thinks that the algorithmically generated rivals do deserve to be taken seriously will not find the local underdetermination problems to be very interesting or important, simply because global

---

[5] For a classic statement of the global underdetermination problem, see Quine (1975).

[6] This radical skeptical scenario was discussed by Wittgenstein (1969) and Russell (1921). See also Acock (1983).

[7] See Melnyk (1997) for an interesting account of what makes one theory a genuine rival of another.

underdetermination is stronger than local. Let us suppose, then, if only for the sake of argument, that philosophers are correct to dismiss the hypothesis that God created the world a mere five minutes ago on the grounds that it is not a genuine rival of any scientific hypothesis about the past. This supposition opens the door for a serious consideration of local underdetermination problems. However, it also means that subsequent conclusions will be conditional upon this being the right response to the global underdetermination problem.

Here is another way to think about this supposition for the sake of the argument: Global underdetermination arguments challenge the very possibility of scientific knowledge, in this case, knowledge of the past. One might think that if we cannot rule out the hypothesis that God created the world five minutes ago, then we cannot reasonably claim to know anything at all about the past. Local underdetermination problems pose a very different sort of challenge to our views about the scope of scientific knowledge. If it were to turn out that local underdetermination problems are especially prevalent in one domain of scientific inquiry, we might naturally begin to wonder whether there are limits or obstructions to our knowledge in that domain. Notice, however, that philosophers who deny that knowledge of the distant past is possible at all – perhaps because they find the global underdetermination argument convincing – will not even notice this second problem concerning the scope of scientific knowledge. In order for this second problem to show up on the radar screen at all, we first need to assume, if only for the sake of argument, that some knowledge of the past is possible.

Condition (b) in my analysis of local underdetermination problems may need some further clarification. Some philosophers have sought to appeal to the non-empirical theoretical virtues in order to block the inference from empirical equivalence to evidential equivalence. If H and H^* are empirically equivalent, it might still be reasonable to prefer H if we could show that it is simpler, or that it has more explanatory power than H^*. This move raises a number of notorious problems, some of which I discuss at greater length in chapter 8: First, how are we to define "simplicity" and "explanatory power" with any precision? Second, why should we suppose that these desirable features are epistemic as opposed to merely pragmatic virtues? What reason is there to think that they are reliable indicators of truth, or approximate truth, or likelihood? (For discussion of some of these problems, see Kukla 1994.) I make no attempt to address these problems here, though I will revisit some of them in chapter 8. It is worth emphasizing, however, that even if these problems could be solved, the appeal to non-empirical theoretical virtue would not necessarily break an evidential tie between empirically equivalent hypotheses, because it is possible neither *H* nor *H** affords a simpler or better explanation than the other. This point is frequently overlooked, because philosophers tend to focus more on global than on local underdetermination problems.

Now for the cases:

(i) *Caytonia* is an extinct gymnosperm from the Mesozoic era. The name was originally given to fossilized reproductive organs consisting of two rows of ovules attached to a central stalk. Two other kinds of structures have since been found. The first are longish pollen-bearing structures. Palynologists suspect that the pollen-bearing structures belong to *Caytonia* because the pollen found in them matches that found in the fossilized ovules. Both reproductive structures are, in addition, associated with clusters of three to six leaflets attached to the end of a stalk. There is no shortage of fossilized parts, and we know that *Caytonia* plants were fairly widespread in Mesozoic North America. The challenge, then, is to infer the architecture of the whole plant on the basis of these fossilized parts. To this day no one knows whether *Caytonia* plants were trees, vines, shrubs, or herbs, and probably no one ever will (Cleal and Tomas 1999, p. 95).

(ii) Ever since the Reverend Edward Hitchock began cataloging and describing fossil footprints in the Connecticut River Valley in the 1830s, vertebrate paleontologists have had the problem of reconciling two distinct taxonomic systems: the familiar system based on skeletal remains and a system of ichnotaxa based on trace fossils, such as footprints. What exactly is the relationship between dinosaur ichnotaxa, such as *Eubrontes* and *Grallator*, and the more familiar taxa that have been identified on the basis of skeletal remains? One problem is that the parataxonomy based on fossil footprints is coarser-grained than the taxonomy based on skeletal remains. Since background theories of taphonomy tell us that the conditions most conducive to the preservation of skeletons in the fossil record are completely different from the conditions most favorable for the preservation of footprints, nearly every fossil trackway poses an underdetermination problem: how can we tell which sort of animal made this particular set of tracks? Footprint fossilization typically happens under the following conditions: A deep pond recedes during a dry season, leaving fine grained bottom sediments exposed to the air. An animal that comes to the pond to drink walks across the muddy flat, leaving a set of footprints. In the days and weeks that follow, the exposed sediment containing the trackway is baked in the sun and hardened. Then at some later point the rains come, the pond is flooded once again, and a new layer of coarser sediment blankets the old, filling in the tracks. If this new layer of sediment hardens in the right way, the footprints will be preserved in the bedding planes of the resulting sedimentary rock. Whereas footprints need to spend some time baking in the sun in order to be preserved, rapid burial, as in a flash flood, is most conducive to the preservation of teeth and bones (Thulborn 1990).

(iii) Jenkins (2000) has criticized the snowball Earth hypothesis, advanced by Kirschvink (1992) and Hoffman et al. (1998) in order to explain evidence of low-latitude glaciation during the neoproterozoic, approximately 800–580 million years ago, by proposing a rival hypothesis that also explains the glacial debris. According to the snowball Earth hypothesis, the entire planet was covered by a layer of ice and snow, on several occasions during the neoproterozoic, for several million years at a time. Jenkins argues that if the earth's obliquity, or the tilt of its axis, had been different during the neoproterozoic than it is today, then low latitudes might have received less energy from the sun than the higher latitudes. Localized glaciation near the equator is just what one would expect to see if the earth's tilt exceeded 54°. The available evidence does not discriminate between the snowball Earth hypothesis of global glaciation and the hypothesis that radical climate changes, including local glaciation at the equator, occurred during the neoproterozoic as a result of changes in the earth's obliquity. Evans (2000) tried to distinguish between these two rival hypotheses by looking for evidence of glacial deposits in regions that would have been near the poles during the neoproterozoic, assuming the high-obliquity hypothesis is correct, but he found none.

(iv) Finally, rival adaptationist hypotheses are all too frequently underdetermined by the available evidence. An adaptationist hypothesis is a hypothesis about what a given trait or behavior is (or was) an adaptation for. For example, Barlow (2000, chapter 1), citing the work of Janzen and Martin (1982), argues that the fruit of the avocado tree is an evolutionary anachronism. The avocado tree co-evolved with the Pleistocene megafauna of Central and South America: gomphotheres, giant ground sloths, and the like. These animals were large enough to swallow avocadoes, carry the pits around in their guts, and later deposit them far from the parent tree. Barlow speculates that the oily green avocado fruit was originally an adaptation for attracting these seed dispersers, all of which have been extinct for approximately eleven thousand years. Since that time, humans have been the main dispersers of avocado seeds. Barlow also points out that Jaguars are known to eat whole avocadoes in the wild, and agoutis gather and bury them just as squirrels bury acorns, but she says that "the fruit of the avocado was not shaped by millions of years of selection for these underabundant, ill-fitted or fickle dispersal agents" (2000, p. 11). But how can we be sure? Was the fruit of the avocado tree an adaptation for attracting ground sloths? Or an adaptation for attracting gomphotheres? While it may be reasonable to suppose that the oily flesh of avocadoes was an adaptation for attracting seed dispersers, hypotheses about *which* fauna were the main dispersers – sloths, jaguars, rodents, gomphotheres, some or all of the above? – are underdetermined by the available evidence.

All four of these cases satisfy condition (a), because in all of them the rival hypotheses are produced by scientists in the course of scientific investigation. They will all count as genuine rivals on any reasonable account of rivalry or theoreticity. The other two conditions deserve a bit more attention.

First, it is relatively uncontroversial that the first two cases satisfy condition (b). Consider the rival hypotheses:

H.   *Caytonia* was a shrub.
H*.   *Caytonia* was a vine.

If anyone were to define "simplicity" or "explanatory power" in such a way as to yield the result that either of these hypotheses is simpler, or affords a more powerful explanation than the other, I think that alone would be good enough reason to reject the proposed definition. Things may not be quite so simple in cases (iii) and (iv). It is conceivable that someone could show that the high obliquity hypothesis is simpler than the snowball Earth hypothesis, or that the snowball Earth hypothesis explains more. In the absence of any precise definition of "simplicity" or "explanatory power," all we can do is to make impressionistic judgments. Sober's (1988) treatment of the notion of cladistic parsimony shows one way in which philosophers of science can give precise definitions of such notions. According to Sober, when a scientist appeals to simplicity to break a tie between competing hypotheses, that appeal is a surrogate for stating a well-confirmed background theory. For example, cladists' appeals to the notion of parsimony are just disguised appeals to background assumptions about the nature of evolutionary processes (1988, pp. 64–65). Anyone who wishes to challenge my claim that these four cases satisfy condition (b) will need to do something analogous to what Sober has done with the notion of cladistic parsimony, and then show, once simplicity has been clearly defined, that the snowball Earth hypothesis (for example) is simpler than the high-obliquity hypothesis.

Do all four cases satisfy condition (c)? Do relevant background theories give us any reason to suspect that the competing hypotheses in these cases are empirically equivalent? I will argue in the next section that the answer is yes.

Before going on to develop the main argument of the chapter, however, I want to address one potential worry about these four examples: None of them are examples of the underdetermination of scientific *theories*. Stanford (2001, S5–6) suggests that the only really convincing examples of empirically equivalent theories come from physics. To be sure, the well-known historical theories of biology and geology (such as Darwin's evolutionary theory, plate tectonics, and so on) are not underdetermined at all in the sense intended here. However, the fact that the four cases I have described are not examples of theoretical underdetermination does not make them any less interesting. Local underdetermination problems such as (i) through (iv) arise during the course of Kuhnian "normal" historical research. The distinction between larger scale theoretical underdetermination and smaller scale underdetermination of hypotheses will not matter for the argument of this chapter.

2.5 HOW HISTORICAL PROCESSES DESTROY INFORMATION

There is one very general reason for thinking that local underdetermination problems are more pervasive in historical than in experimental science. Background theories of geology, and especially taphonomy, tell us that many historical processes – the fossilization process, the processes of weathering and erosion, continental drift, subduction, glaciation, and so on – are *information-destroying processes*, rather like housecleaning and document shredding. Elliott Sober (1988, pp. 3–5) uses the following example to illustrate this concept of an information-destroying process. Suppose a person releases a ball from the rim of a giant bowl. A later observer happens along and finds the ball resting at the bottom of the bowl. It will be impossible for the observer to infer from which point along the rim the ball was released.[8] No one hypothesis about the point of release is any more probable than another. In this case, rival hypotheses about the point of release are underdetermined by the observable evidence, because all of them are empirically equivalent. The interesting thing about the example, however, is that we have background knowledge (of bowls, gravity, and so forth) that leads us to expect that rival hypotheses will be empirically equivalent in the strong sense. We can even explain how and why the process by which the ball rolls to the bottom of the bowl destroys information about the point from which it was released.

The situation in prototypical historical science is analogous. Kemp (1999) describes various kinds of incompleteness in the fossil record: For instance, *biogeographic incompleteness* is a serious problem for paleoecologists. Suppose that a population of terrestrial animals migrates seasonally between dry upland areas and wetter lowland areas that are well drained by rivers. Conditions in dry upland areas are not well suited to fossilization, so it is a good bet that the only members of this species who make it into the fossil record will be the ones that die in lowland areas, along river banks or in floods. This means, however, that the fossil record will give us a distorted picture of the range of these animals. Another upshot of this is that some biological communities will be far more extensively represented in the fossil record than others. The point is simply that we know that the fossilization process destroys information about biological communities in dry upland areas. Another relevant problem discussed by Kemp is that of *stratigraphic incompleteness*, which arises because sediments do not accumulate at a constant rate. The periodic flooding of a major river, such as the Mississippi, affords a good example of this. Since more sediment is deposited during floods than at other times, when scientists look at a layer of sedimentary rock, they are looking at sediments that accumulated in fits and starts. Kemp points out that if there were a period during which no sediment was deposited, that can have a distorting effect on the fossil record. Suppose that some population of organisms living in the neighborhood was evolving at a steady rate during a given stretch of time. Suppose, further, that a large amount of sediment was deposited during the early part of this stretch, and a lot of sediment was deposited during the later part, with a lengthy gap in between, during

---

[8] Interestingly, this is also a good illustration of Ben-Menahem's (1997) conception of historical necessity. According to her, events are necessary when they are relatively insensitive to initial conditions, and contingent when they are relatively sensitive to initial conditions.

which time the local rivers, for whatever reason, happened not to flood. This stratigraphic incompleteness will create the illusion of rapid, or even punctuated evolutionary change in the population. Scientists looking at the record will not be able to discriminate between the hypothesis of gradual evolutionary change during a time in which no sediment accumulated, or the hypothesis of steady sedimentation and rapid evolutionary change.

I conclude that condition (c) is satisfied in all four of the non-trivial cases described above, which means that they are all bona fide cases of local underdetermination. In all four cases, our background knowledge of the incompleteness of the geological record gives us at least some reason to think that the rival hypotheses are empirically equivalent.

Laudan and Leplin (1991) suggest that there might also be some general reasons for thinking that H and H^*, though consistent with all the observations made so far, are not really empirically equivalent. First, the empirical consequence class of any hypothesis is determined, in part, by auxiliary assumptions that are subject to revision over time. It is at least possible that paleontologists will revise some of the background assumptions of taphonomy – the very background theories that, for the moment, give us reason to think that H and H^* are empirically equivalent – and if this were to happen, it could turn out that H and H^* are not empirically equivalent at all. This point is well taken, and it is one reason why we should be careful about jumping to the conclusion that any pair of rivals, H and H^*, are empirically equivalent. However, there is no reason to think that our background theories of taphonomy are going to be significantly revised anytime soon, and since those background theories do provide some reason for thinking that the rival hypotheses in these four cases are empirically equivalent, it is correct to describe these as cases of local epistemic underdetermination. It is also worth pointing out that Laudan and Leplin's main target is the global underdetermination argument; their observation about the instability of auxiliary assumptions is compatible with the existence of local underdetermination problems, such as I have described.

Laudan and Leplin (1991) also emphasize that the range of the observable is liable to change, which is another reason why we should hesitate to conclude that any two theories or hypotheses really are empirically equivalent. This point, too, is well taken. Suppose that engineers devise a new fossil detection gizmo that enables scientist to study fossils buried in places that are otherwise inaccessible. It is possible that the new gizmo would enable scientists to find a "smoking gun" that would discriminate between the hypothesis that *Caytonia* was a vine and the rival hypothesis that it was a shrub. But this sort of consideration is not terribly helpful. Scientists have found loads of partial *Caytonia* fossils, suggesting that the conditions favorable to fossilization of the leaves and reproductive structures were, for whatever reason, unfavorable to the preservation of the other parts of the plant. Based on this track record, there is some reason to doubt that the smoking gun is even out there for us to find.

In sum, Laudan and Leplin's arguments show that it would be rash to assert that the rival hypotheses in these four cases are empirically equivalent, but that is not what (c) asserts. (c) only says that background theories about historical processes lend some support to the claim that *H* and *H** are empirically equivalent.[9]

2.6 A FOSSILIZED DINOSAUR HEART

What about cases in which, contrary to what our background theories may lead us to believe, someone *does* find a smoking gun that discriminates among hypotheses that once looked to be empirically equivalent? Consider, for instance, the question whether or not dinosaurs were endothermic. A few decades ago, it might have been reasonable, given the available background theories, for scientists to conclude that no one will ever find a smoking gun to lend support to one or the other hypothesis. In other words, a few decades ago, hypotheses about dinosaur metabolism may well have constituted a local underdetermination problem, according to the above analysis. However, in recent years, scientists have discovered a number of different historical traces – including an apparent fossilized dinosaur heart that has four chambers and one aorta, just like a bird's – that clearly support the hypothesis that dinosaurs were endothermic (Fisher et al. 2000). This example seems to show that not all local underdetermination problems in historical science are permanent.[10]

Yet there are two reasons for thinking that this serendipitous smoking gun is one of those exceptions that proves the rule. First, some respected

---

[9] Compare also Deborah Mayo's account of severe testing (1991; 1997). A severe test might be another way of breaking an evidential tie between rivals that are consistent with all the observations so far. However, severe testing (in Mayo's sense) is only possible in an experimental setting.

[10] For another fascinating example that remains controversial, see Schweitzer, Wittmeyer, Horner, and Toporski (2005), who report having discovered soft blood vessel tissue inside a bone from *Tyrannosaurus rex*.

scientists have doubted that the object which Fisher and colleagues studied using CT scans is a fossilized heart at all. Rowe, McBride, and Sereno (2001) point out that the alleged fossilized heart was found inside the chest cavity of a *Thescelosaurus* sekeleton, in the sandstones of the Hell Creek Formation, in Montana. They argue on the basis of taphonomy that it is highly implausible to suppose that the internal organ of a dinosaur could ever have been preserved in such sedimentary environments. They suggest that Fisher and colleagues were in fact looking at an ironstone concretion and not at a fossil at all, for "ironstone concretions are notorious for producing suggesting and misleading shapes," and they have often been found in conjunction with dinosaur bones in the American west (Rowe et al. 2001, p. 783a). Regardless of the eventual outcome of this debate, it is instructive to see that in this case, specialists are arguing from background theories about information-destroying processes to the conclusion that what seems like a stunning example of a smoking gun may not be a smoking gun at all. The background theories of taphonomy are that powerful.

Second, suppose that the object scanned by Fisher et al. really is a fossilized dinosaur heart, and that it is a serendipitous smoking gun. If so, that only gives rise to new research questions, and – arguably – new local underdetermination problems. For example, Fisher and colleagues point out that *Thescelosaurus* is an ornithischian ("bird-hipped") dinosaur, whereas modern birds, with their four-chambered hearts, are thought to be more closely related to the saurischian ("lizard-hipped") dinosaurs. Did the four-chambered heart evolve independently in several dinosaur lineages, or did it evolve early on in dinosaur history, perhaps even before the saurischians and ornithischians diverged? This, as Fisher and colleagues point out, "remains an open question." Since there is no reason to expect that we will find any more fossilized dinosaur hearts, the answers to such questions about the evolution of dinosaur hearts will probably remain locally underdetermined. Thus, even if it is a genuine smoking gun, the fossilized dinosaur heart only gives rise to new local underdetermination problems.

2.7 THE ROLES OF BACKGROUND THEORIES IN HISTORICAL *VS.* EXPERIMENTAL SCIENCE

Permanent local underdetermination problems are widespread in historical science, but less common in experimental science. Why? The main reason for this has to do with the different roles that background theories play in historical *vs.* experimental science. In historical science, as I hope to have shown, background theories about information-destroying historical processes lead to widespread local underdetermination problems, because they mean that condition (c) in the above analysis of such problems will very often be satisfied. Such background theories imply that there are a great many things that scientists will never know about the distant past. Or to put it another way, those background theories serve (or should serve) as a check to the epistemic ambitions of historical researchers. Although they can develop new technologies for identifying and studying potential smoking guns, such as the CT scans used by Fisher and colleagues to study the internal structure of the alleged dinosaur heart, historical scientists can never manufacture a smoking gun. If, in fact, every single dinosaur heart was destroyed by the fossilization process, there is nothing anyone can do about it.

On the other hand, background theories play a very different role in experimental science. Whereas background theories about information destroying processes must dampen the epistemic ambitions of historical scientists, a different set of background theories serves (and should serve) to enlarge the epistemic ambitions of experimentalists. Scientific realists, such as Richard Boyd (1985), have long emphasized the dialectical relationship between theory and method in experimental science. Experimental design always depends heavily upon background theories that tell scientists how to build experimental apparatus that will enable them to manipulate certain test conditions. This experimental manipulation then gives them a way to test new theories and hypotheses which, if confirmed, may provide new clues for the design of future experiments. In experimental science, background theories serve as guides for the design of new experiments whose purpose is to produce results that evidentially discriminate between rival hypotheses.

To illustrate this point, consider a piece of highly publicized recent experimental research on pheromones in German cockroaches (Nojima et al. 2005). Scientists already knew, based on earlier work, that female cockroaches give off a pheromone that can attract males from some distance away. And they knew that if they could isolate the pheromone, the discovery would have commercial applications. What better way to lure cockroaches into bait stations? However, female cockroaches produce such tiny quantities of the pheromone that the researchers had to remove the relevant glands from 15,000 individuals in order to produce enough substance to work with in the lab. Even then, the substance extracted from the glands is chemically complex. So researchers had to find some way of breaking the substance down into its chemical components in order to determine which component is the sought-after pheromone. They used a gas chromatograph to identify the various components of the extract. Then they hooked up a detached cockroach antenna to a set of electrodes and exposed the antenna to each of the chemical components. When the antenna sent an electric charge through the electrodes, the scientists reasoned that they had discovered the pheromone. The pheromone was a previously undiscovered compound, which they named blatellaquinone.

In this case (and one could easily multiply examples), background theories showed the scientists how to create new evidence in the lab. Here the new evidence was the electric charge sent by the antenna. The background theories included well-established theories about the chemical structures of pheromones, theories about gas chromatography, and theories about the behavior of cockroach antennae. These theories together served to enhance rather than dampen down scientific ambitions. Moreover, to confirm that they really had identified the pheromone, the scientists created a synthetic version of blatellaquinone and tried it out on cockroaches – an example of how manipulation of the tiny can help eliminate underdetermination. The synthetic pheromone had the same effect on male cockroaches as the original extract from the females' glands.

Here, then, is the central argument of this chapter:

P1.   In prototypical historical science, background theories tell us how historical processes destroy information. Background theories do not usually play this *dampening role* in experimental science.

P2.   In classical experimental science, by contrast, one of the main functions of background theories is to serve as guides for the design of new experimental apparatus whose purpose is to produce new evidence that breaks evidential ties between seemingly empirically equivalent hypotheses. They do not usually play this *enlarging role* in historical science.

C1.   Hence, there is a good reason for thinking that local underdetermination problems will be more widespread in historical than in experimental science.

C2.   Hence, prototypical historical science is, in one sense, epistemically disadvantaged relative to classical experimental science.

Notice that C2 is just a restatement of one of the main claims of chapter 1 – namely, that there is an epistemic scope asymmetry between the past and the tiny. There I claimed that we can know more about the tiny than about the prehistoric. It is now possible to say with greater precision than before just what this means: Since every local underdetermination problem amounts to a gap in scientific knowledge, and since permanent local underdetermination is more pervasive in historical than in experimental science, there will be more permanent gaps in our knowledge of the past than in our knowledge of the tiny. Once again, we have traced the epistemic asymmetry between the past and the tiny to the deeper asymmetry of background theories. In historical science, the background theories tell us *how nature has destroyed the evidence*. In experimental science, they tell us *how to make new evidence*.

It is worth emphasizing, in conclusion to this chapter, that I do not mean to say that any particular historical theory, such as Darwin's evolutionary theory, is less well confirmed than any particular theory about the microphysical world, such as quantum theory; or that prototypical historical science is in any way less scientific than classical experimental science; or that we do not really have scientific knowledge of the past; or that historical science is less worth doing than experimental science. All that I claim to have shown is that there are limits to our scientific knowledge of the past that do not similarly limit our knowledge of the tiny. This is a skeptical conclusion, but it is an example of *mitigated* skepticism. The argument of this chapter actually presupposes that we already have quite a lot of background knowledge of the past – something which a thoroughgoing skeptic would deny.

Finally, some of Cleland's most important insights are actually compatible with this modestly skeptical result. For example, she justly infers from the fact that historical scientists and experimental scientists exploit different aspects of the asymmetry of overdermination, that "neither practice may be held up as more objective or rational than the other" (2002, p. 476). Everything that I have said about the scope asymmetry between the past and the tiny is compatible with the idea that historical and experimental science are about equally objective and rational.

3

Manipulation matters

In this chapter, I turn from questions about the scope of scientific knowledge to questions about the possibility of scientific knowledge. I also turn from the role asymmetry of background theories, which was the main theme of chapter 2, to the asymmetry of manipulability. What are the main arguments that philosophers have used to try to establish the possibility of scientific knowledge of unobservables? And how does the asymmetry of manipulability affect those arguments?

Let us say that *minimal epistemic realism* with respect to some kind *K* of unobservables is the view that it is possible to have scientific knowledge of unobservables of that kind.[1] In this chapter, I will argue that past events, entities, and processes, on the one hand, and tiny events entities and processes, on the other, constitute two kinds of unobservables. There is a relevant difference between these two kinds of unobservables that has a bearing on the arguments for minimal epistemic realism: We cannot manipulate the past, but we can often manipulate the tiny. This difference has the consequence that the standard arguments for minimal epistemic realism give less support to minimal epistemic realism with respect to one kind of unobservables, than to realism with respect to the other. Thus, the argument of this chapter will show that one surprising consequence of the asymmetry of manipulability is *historical hypo-realism*: The standard arguments for realism (if they are any good at all) give less support to minimal epistemic realism about the past than to minimal epistemic realism about the tiny.

The first step in developing this line of argument is to make plausible the idea that the past and the tiny are two kinds of unobservables. Are

---

[1] This is meant to be equivalent to the definition offered by Leplin (1997, pp. 102–103), though he does not relativize the definition to kinds of unobservables.

dinosaurs really unobservable, though? One might think that dinosaurs are observable, on the grounds that there are certain subjunctive conditional statements that are true of dinosaurs but not true of other things, such as electrons. For example, it is true that if there were a dinosaur out on the green, then I would be able to see it, smell it, etc., but this is not at all true of electrons. If the claim that *x* is observable just means that some such subjunctive conditional statements are true of *x*, then there is nothing wrong with saying that dinosaurs are observable (in one sense of "observable"). However, this does not change the fact that we cannot observe them, because they no longer exist. Dinosaurs were observable, but they are no longer observable. Dinosaurs may even be observable in a sense in which electrons are not observable. But we can just as well say that electrons are observable in a sense in which dinosaurs are not – for electrons exist now.[2]

The distinction between things that are unobservable in virtue of their small size and things that are unobservable because they existed or occurred in the past is neither exclusive nor exhaustive (Carman 2005). It is not exclusive because many tiny things, such as the electrons of the dinosaurs, also existed in the past. It is not exhaustive because some presently existing things, such as the center of the earth, are inaccessible to observation for reasons having nothing to do with their size. Nevertheless, it seems possible to classify unobservables according to what makes them unobservable. And facts about what makes things unobservable can be relevant to questions about the possibility of scientific knowledge of them.

3.1 CAN WE OBSERVE THE PAST?

Everyone has heard the factoid that it takes light approximately eight minutes to reach the earth from the sun. Suppose that the sun is suddenly annihilated. About eight minutes go by before anyone on Earth notices that anything has happened. During that time, someone looks up at the sky in the direction of the sun. Does she observe the sun? No matter how we choose to answer that question, we will have to face some strange consequences. If we say yes, she observes the sun, then we will be forced to say that it is possible to observe something that no longer exists. On the other hand, if we say no, she does not observe the sun, but that she only

---

[2] I thank Michael Lynch for helping me to see how to put this point most effectively.

observes the sun's light (or a bright yellow patch, or anything other than the sun), then we will be forced to say that no one has ever observed the sun. Either answer is weird. Either we can observe things that no longer exist, or else the sun is an unobservable entity. Even stranger still, this result generalizes to things that are much closer to us than the sun.

The past and the distant are related in a special way. The more powerful our telescopes, the more distant the stars we can observe. The more distant the stars, the longer it takes light traveling from them to reach our telescopes. Some astronomers are now hoping to build a telescope on the moon that will be able to detect light that has been traveling for billions of years. Although it sounds paradoxical, one way in which astronomers can learn more about the distant history of the universe is to keep building ever more powerful telescopes that can detect light traveling from ever more distant stars. Someone might well think that scientists are observing the past when they look through their telescopes at very distant star that was extinguished a long time ago. If we can observe things that have not existed for many millions of years, what becomes of the original distinction between things that are unobservable because they existed or occurred in the past, and things that are unobservable owing to their small size?

In response to the line of argument just sketched, one must draw the distinction between the two kinds of unobservables in a slightly different way. According to the view that we are considering, having existed or occurred long ago is not by itself sufficient to render something unobservable. We would need to include distance as well as time in our account of what makes, say, the dinosaurs unobservable. They cannot be observed because they existed long ago in a nearby region of the universe. Even if we were to decide that it makes sense to say that we can observe some past things and events, we could still distinguish between (A) things and events that are unobservable owing to their small size relative to us, and (B) things and events that are unobservable owing to their distance in time and nearness in space relative to us.

In his recent book on philosophical issues in archaeology, Peter Kosso (2001) gives a rather different argument for the claim that we can observe the past. Kosso's main goal is to answer the charge that history and archaeology have a special epistemic handicap that distinguishes them from the natural sciences:

> The challenge for historical studies is often distinguished from that of natural science by an alleged observational disadvantage for the historian. The

intuition behind this distinction has it that since all of the objects and events of interest to the historian or archaeologist are dead and gone, they are not amenable to observation. There is nothing equivalent to a telescope or a microscope with which a viewer of the twenty-first century can bring into focus an image of the Persian wars. (Kosso 2001, p. 40)

Kosso's own response to this argument is to try to equalize things by showing that we can observe the past: "Being in the past is neither in principle nor contingently an impediment to an object's being observed" (Kosso 2001, p. 40). If he is right about this, that would pose another challenge to my distinction between two kinds of unobservables. But Kosso's argument is unconvincing.

To begin with, notice that our earlier reflections about the observation of stars that have not existed for many millions of years assumed that light is the medium by which information is transmitted to observers. The reason why we cannot observe dinosaurs is that this signal travels way too fast; the light reflected back into space from the last living dinosaur is now far, far away. Kosso then argues that there is no reason in principle why observation could not involve informational signals that travel much more slowly.

> While there is an upper limit on the speed of transmission of information, namely the speed of light, there is no lower limit. Furthermore, slower signals mean that closer objects can be observed as they were in the more distant past. (Kosso 2001, p. 43)

Suppose that in addition to light, the sun emitted another far slower informational signal that we could detect with some other sensory modality. Suppose this other-than-light signal travels at exactly one tenth the speed of light. When we see the sun, we are in fact seeing an object that existed approximately eight minutes ago, but when we observe the sun via this other sensory modality, we perceive an object that existed approximately eighty minutes ago. Other things being equal, the slower the signal, the further back in time we can observe. Here, then, might be a completely new reason to think that we can observe the past: if we can observe things via informational signals that travel much more slowly than light, then perhaps we can observe (in some sense) Lincoln's delivery of the Gettysburg address. Maybe we could even observe dinosaurs. Kosso, however, veers away from these implausible claims, saying, "I am not going to advocate that we say that we in the twenty-first century can see Napoleon or that we can observe the Battle of Hastings or the murder of Alcibiades"

(Kosso 2001, p. 49). Why not advocate saying these things, if (as Kosso holds) being in the past is no impediment to being observed?

Kosso's view concerning the relationship between observation and informational signals is somewhat confusing. He writes that "observation is more than stimulus; it is the reception of information" (Kosso 2001, p. 45). Suppose that Smith is adrift at sea in a life raft. Before dying in a storm, Smith places a message in a bottle and releases it into the current. This message is an extremely slow traveling informational signal. One year later, long after Smith has perished, Jones discovers the bottle and reads the message, which describes Smith's plight on board the raft. No one would say that Jones has thereby observed Smith. Rather, Jones has observed the message and drawn some inferences based on that evidence. He does not observe that Smith's raft was yellow; rather, he infers that the raft was yellow, based on Smith's testimony. The point is that all of our evidence about the past comes in the form of informational traces or signals, but receiving an informational signal from something is not sufficient for observing it, or for observing that it has some property.

There are clear-cut cases of entities, events, and processes that we will probably never observe, even with technological aids, because they are so tiny. In addition, there are clear-cut cases of entities, events, and processes that we will probably never observe because they existed or occurred in the past. Thus, there are two kinds of unobservables. This is the first premise of my argument in this chapter.

3.2 THE CONTEXT-DEPENDENCE OF THE RANGE OF THE OBSERVABLE

It is impossible to draw any precise boundary between what we can observe and what we cannot. The boundary is vague and shifting, so that the answer to the question, "Can x be observed?" will often be that it depends. For example, whether a particular star can be observed on a particular night will depend on our position on the earth's surface, local weather conditions, the quality of our telescope, and much else.

In a now classic paper on scientific realism, Grover Maxwell drove this point home in a vivid way:

There is, in principle, a continuous series beginning with looking through a vacuum and containing these as members: looking through a windowpane, looking through glasses, looking through binoculars, looking through a low-power microscope, looking through a high-power microscope, etc. in the order given. The important consequence is that, so far, we are left without criteria which would enable us to draw a non-arbitrary line between "observation" and "theory". (Maxwell 1962, pp. 7–8)

Theory enters into the picture here because the technological devices Maxwell mentions were all designed and manufactured with the help of theories of optics. Some things that were once unobservable – such as, say, microbes or the moons of Jupiter – became observable with the development of new technological aids to observation. But if we recall the special relationship between the past and the distant, it turns out that Maxwell's point also applies to observation of the past. Better telescopes enable us to see things that are further and further away, which (in a sense) enables us to see further into the past.

I mention Maxwell's argument here in order to fend off one potential objection. One could argue that since there is no non-arbitrary way of distinguishing the observable from the unobservable, there is no point in trying to distinguish kinds of unobservables on the basis of different facts about them that render them unobservable. A simple analogy will help to show what is wrong with this objection. Maxwell's argument shows at most that observability is a vague predicate, rather like baldness (van Fraassen 1980). Most people would agree that we are left without criteria which would enable us to draw a non-arbitrary line between "baldness" and "hairiness." However, even after granting this point, we can identify clear-cut examples of bald people. Not only that, but we can distinguish among different kinds of bald persons according to the causes of their baldness. Some people, for example, have lost their hair naturally, over the course of many years, whereas others have paid someone else to shave their heads. This, roughly, is how I propose to approach the problem of unobservables: begin with relatively uncontroversial examples of unobservable entities, events, and processes, and then classify them according to the facts about them that make them unobservable.

3.3 TWO SPECIES OF SCIENTIFIC REALISM

Since there are two kinds of unobservable entities, we can also distinguish between two species of scientific realism, where realism is construed as a view about the possibility of scientific knowledge of unobservables.

Historical Realism (minimal, epistemic). We can and do have some scientific knowledge of things and events that are unobservable because they existed or occurred in the past.

Experimental Realism (minimal, epistemic). We can and do have some scientific knowledge of presently existing and occurring things and events that are unobservable owing to their small size relative to us.

Notice that these are species of a genus, which we might call generic realism.

Weak, or Disjunctive Generic Realism (minimal, epistemic). We can and do have some scientific knowledge of unobservable things and events.

This generic realism is implied by historical realism and also by experimental realism. In other words, if a realist philosopher of science wanted to argue for generic realism, it would be sufficient for her to provide a cogent argument for either species of realism. One can validly infer generic realism, in this first sense, from either species of realism without committing any hasty generalization. This is disjunctive realism because the idea is that we have knowledge of unobservables of kind K, or of unobservables of kind $K1, \ldots$ or of unobservables of kind Kn.

Strong, or Conjunctive Generic Realism (minimal, epistemic). We can and do have some scientific knowledge of unobservable things and events, including those that existed or occurred in the past, as well as those that are too small to be observed.

This second variety of generic realism is much stronger than the first. In order to give support to this second variety generic realism, it would be necessary to argue for both historical and experimental realism. Anyone who produced an argument for experimental realism and claimed that she had thereby given support to strong generic realism would have committed the fallacy of hasty generalization.

It is not always clear whether philosophers of science are in fact guilty of committing this hasty generalization. The trouble is that many current statements of the realist position are ambiguous with respect to the weak and the strong versions of generic realism. This is because they do not distinguish between the different kinds of unobservables, or between the different species of realism. If philosophers intend only to assert the weak version of generic realism, then there is no problem at all. However, if they intend to assert the strong version, there is serious danger of hasty generalization.

Of course, the inference from one species of scientific realism to strong generic realism would be especially problematic if relevant differences between the kinds of unobservables meant that the case for one species of realism is stronger than the case for the other. I will argue in this chapter that assuming the abductive arguments for realism are any good at all, the positive case for historical realism is weaker than the case for experimental realism.[3] The idea is that there is an asymmetry built into the arguments for scientific realism; for lack of a better term, I will call this *historical hypo-realism*. This asymmetry in the arguments for realism follows from the asymmetry of manipulability.

How might one go about arguing for the thesis that we can and do have knowledge of unobservables of a given kind? Most realists proceed by giving arguments to the best explanation, or (to use C. S. Peirce's term) abductive arguments. Such arguments have the following structure: A number of hypotheses, $H1, H2, \ldots$, would, if true, explain some puzzling fact or phenomenon. Of these, $H1$ is the best. Therefore, $H1$ is probably true. In this chapter, I examine two versions of this abductive strategy: the basic abductive argument for realism and the more complicated inference to the best explanation of the empirical success of science. One consequence of the asymmetry of manipulability is that these arguments give less support to historical realism than to experimental realism.

Before moving on, it is worth pausing to consider two important objections. To begin with, one might think that historical realism has a higher degree of initial plausibility than experimental realism does. If that were the case, then the arguments for historical realism would not need to be as good as those for experimental realism. Someone who takes this line will need to point to a relevant difference between the two kinds of realism that explains why the one is more plausible than the other. Yet even if the one species of realism did have a higher degree of initial plausibility than the other, it would still be worth asking whether the arguments for the two species of realism have the same degree of force. Intuitively, one might think that the species of realism that is alleged to have the greater initial

---

[3] Actually, it would be problematic even if the cases for the two species of realism were equally strong. In order to see why, consider what it would take to argue for a conjunctive claim, P and Q, where the probability of P is 0.7 and the probability of Q is 0.7, too. The probability of P and Q will be 0.49. It would be a mistake to assume that since P alone is well supported, P and Q must have the same degree of support. The problem becomes even worse if the independent probabilities of the conjuncts differ. Suppose the probability of P is 0.7, while the probability of Q is 0.3. Then the probability of P and Q will be 0.21. So someone who supposes, from the fact that P is well supported, that P and Q is just as well supported, will be making an even more egregious mistake.

plausibility would also be better supported by the arguments. It would be interesting to learn that the very opposite is the case, and that the species of realism with the lower degree of initial plausibility is actually better supported by the arguments.

Second, a number of philosophers of science have sought to show that the abductive arguments for realism are no good at all. Bas van Fraassen, Larry Laudan, and others have raised some powerful blanket objections to this realist strategy. Van Fraassen, for example, zeroes in on the question of what standard we shall use to determine which among a pool of competing potential explanations is the best. Suppose we use simplicity as the standard. Even if we could define simplicity in a way that would enable us to tell with precision which hypothesis is the simplest, we would have to ask whether there is any empirical evidence to support the claim that the simplest explanation is likeliest to be true. Why assume that the world we live in is simple? (This "problem of abduction" parallels David Hume's famous problem of induction. Hume asked whether there is any empirical evidence to support the claim that the future will resemble the past.) If, as van Fraassen has argued, inference to the best explanation is not to be trusted, then the abductive arguments for realism are not to be trusted, either.

Another widely discussed blanket objection to the abductive arguments for realism comes from Larry Laudan (1981), who argues that the realist case is undermined by historical examples of instrumentally reliable theories whose central theoretical terms did not refer. These examples, such as the phlogiston theory of combustion and the caloric theory of heat, show that theories that are completely wrong in what they say about the unobservable world can still enjoy tremendous empirical success. We cannot explain the success of those theories by saying (as scientific realists would like to say) that they are true or approximately true. However, as soon as we admit that some other explanation of empirical success (aside from truth or approximate truth) is the best explanation in some cases, why not suppose that it is the best explanation in other cases?

If these blanket objections succeed, they show that the abductive arguments for realism do nothing to support either historical or experimental realism. In that case, it would be false that the arguments lend more support to one species of realism than to the other, for they would lend no support to either species of realism. For that reason, it is necessary to conditionalize the thesis of this chapter: if the abductive arguments for realism are any good at all – that is, if realists can give good answers to the blanket objections to those arguments – then they give less support to historical than to experimental realism.

3.4 TWO ROLES FOR UNOBSERVABLES

The linchpin of my argument is the idea that scientists posit unobservable entities and mechanisms for different purposes. Or to put it another way: unobservable entities and mechanisms can play different roles, or perform different functions in science. For example, some unobservable entities and mechanisms function as *unifiers* of the phenomena: by supposing that those unobservable entities and events exist and occur, scientists can give a more or less unified or coherent account of the observable evidence. In science, however, unobservable entities and mechanisms sometimes also function as *tools for producing new phenomena*. Laboratory scientists often use elaborate experimental apparatus to produce new observable results by manipulating things that we cannot observe. Thus, it makes sense to distinguish between the *unifying role* and the *producing role*.[4] Some unobservable entities and mechanisms play both roles simultaneously. But some unobservables can only play the unifying role, and never the producing role.

Some might object to this distinction between the unifying role and the producing role, on the following grounds: the supposition that certain kinds of entities really do exist, and that scientists really are interacting with them when they perform certain experimental manipulations sounds like a realist one. This realist supposition needs to be defended, presumably with the help of the standard abductive arguments for realism. But this in turn means that the supposition that anything does play the producing role must be defended by appeal to considerations having to do with explanatory or unifying power. So the distinction between the producing and the unifying roles threatens to collapse.[5]

In order to address this worry, it will help to pull apart two quite different issues: (a) the issue of whether the distinction between the producing

---

[4] This distinction between the unifying role and the producing role echoes Robert Nola's (2002) distinction between "realism by hypothesis" and "realism by manipulation." Realism by hypothesis is realism with respect to unobservables that play the unifying role, whereas realism by manipulation is realism with respect to unobservables that play the producing role. What is missing from Nola's discussion, however, is any consideration of historical science.

[5] I thank an anonymous reviewer for raising this objection.

role and the unifying role makes any sense to begin with, and (b) the issue of whether anything in fact does play either of these two roles. One way to see that the distinction does make sense is to shift the focus momentarily from unobservables to unobserved observables. Things that are readily observable but not currently observed can play either of these roles. For example, a deer in the woods that no one is currently observing could serve to unify various phenomena (tracks in the snow, bushes stripped of their leaves). But no one is using the deer as a tool to produce new phenomena. On the other hand, when you send an email message to a colleague, you use the colleague's computer as a tool for producing new phenomena (i.e. as a tool for causing your colleague to have certain visual experiences). In the present context you are not using your colleague's computer to unify any phenomena, although you could do that in a different context. Since it is clear that unobserved observables can play either of these two functional roles, the distinction surely makes sense when applied to unobservables, too.

An example will help to illustrate this distinction between the unifying role and the producing role. The paleontologists Kevin Padian and Paul Olsen (1989) once devised an experiment to test rival hypotheses about the posture of therapod dinosaurs, a group of animals that has been extinct for 65 million years. They wanted to determine whether therapods walked with a forward-leaning semi-erect posture, or with a fully erect posture. The experiment they devised was based on their earlier research on ancient crocodilians (Padian and Olsen 1984). They had coaxed a living crocodile to walk through a patch of mud using the two different crocodile gaits – the high walk and the low-slung sprawling walk. They had then compared these tracks to the preserved tracks made by ancient crocodilians. Finally, they had inferred, on the basis of the observable similarities between the two sets of tracks, that the stances and gaits of the ancient crocodilians resembled those of the living animals. Padian and Olsen argue that inferences like this will go through, so long as we are comparing ancient and living organisms belonging to "a single phylogenetically restricted group" (1989, p. 231). The underlying methodological assumption is that "functional similarity between animals is correlated with degree of phylogenetic relationship" (1989, p. 232). Since birds are probably the closest living relatives of the therapod dinosaurs, Padian and Olsen ran a similar experiment with a 25 kg South American rhea from the Oakland Zoo. They had the bird walk across a bed of potter's clay, and they observed some similarities between the rhea's tracks and those of ancient therapods: both animals placed their feet near to the midline

71

of the trackway; in both cases, the middle toe is pointed slightly inward; neither animal's toes leave drag marks, and so on. These similarities, they argue, lend some support to the hypothesis that therapods had an erect posture, like living ratites.

In this example, Padian and Olsen are using the living (and observable) rhea as a tool for producing new phenomena – namely, the tracks in the bed of clay. Whereas the living bird plays the producing role, the extinct (and hence unobservable) therapod dinosaurs play the unifying role only. The scientists cannot manipulate those dinosaurs to produce any new experimental results. However, by supposing that the animals existed so many millions of years ago, they can unify the observable evidence – namely, the fossilized tracks and skeletal remains.

There is also a rich tradition of experimentation in archeology, none of which involves actual manipulation of the past. Coles (1973, pp. 27–34) provides vivid descriptions of experiments in which archaeologists created replicas of ancient ploughs, hitched them up to oxen, and used them to work a number of test plots. The aim of these experiments was to test various hypotheses about how long it would have taken ancient farmers to plough a given acreage, how frequently they would have had to repair their tools, and much else.

3.5 TWO BASIC ARGUMENTS FOR REALISM: DEVITT AND HACKING

I begin with a version of what Hacking calls "the experimental argument for realism" (1983, p. 265), and Nola calls "the argument from manipulability" (2002). It goes like this:

P1.   Scientists can interact with what they take to be unobservable x's (e.g. electrons and positrons) and thereby alter observable conditions in predictable and systematic ways.

P2.   The fact that scientists are thus able to control the observable by means of what they take to be the unobservable x's would be inexplicable if those x's were not real, or if they lacked the properties scientists take them to have.

C   Therefore, the unobservable x's are probably real, and probably have the properties that scientists take them to have.[6]

---

[6] One reviewer has suggested to me that Hacking never intended to give an explicit experimental argument for realism. Either he is just being rhetorical, or else he is merely doing autobiography, and explaining what sorts of considerations in fact led him to become a

Leaving aside the question of how powerfully this argument supports experimental realism (minimal, epistemic), it clearly does not support historical realism at all. Our experimental interventions may give us some reason to think that electrons are real – and no reason for thinking that N-rays, gemmules, and angels are. However, no such interventions will justify the belief that mastodons, as opposed to one-eyed giants, were real. Scientists also believe that therapod dinosaurs were real while *Brontozoa* – the humongous, flightless, Mesozoic poultry that the Reverend Edward Hitchcock posited in order to explain the largest of the footprints he found in the Connecticut River Valley – were not (Hitchcock 1858/1974, pp. 178–179). But the argument from manipulability contributes nothing to the justification of that belief. Since the experimental argument for realism is unavailable in this context, the cumulative case for minimal epistemic realism with respect to historical science is going to be somewhat weaker than the case for realism with respect to experimental science.

A more promising argument for historical realism is the stripped-down inference to the best explanation that Michael Devitt has called the "basic" abductive argument for realism, in order to distinguish it from the so-called inference to the best explanation of the success of science. Here is Devitt's argument:

> The basic argument for the unobservable entities is simple. By supposing they exist, we can give good explanations of the behavior and characteristics of observed entities, behavior and characteristics which would otherwise remain completely inexplicable. Furthermore, such a supposition leads to predictions about observables which are well-confirmed; the supposition is "observationally successful". Abduction thus takes us from hypotheses about the observed world to hypotheses about the unobservable one. (Devitt 1991, pp. 108–109)

Elsewhere Devitt stresses that whether or not it explains the empirical success of science, realism itself is observationally successful (1991, p. 114).[7] I have just two observations to make about this argument: First, it is available to those who want to defend realism with respect to historical science, so the case for historical realist epistemology is at least

realist. This may be true. Nevertheless, I think that attributing some such argument to Hacking may be the most charitable way of reading him. My argument in this chapter does not depend on whether this is the correct reading of Hacking; the fact that one can construct a simple (abductive) experimental argument for realism is all that I need.

[7] See Hanen and Kelley (1989) for an illustration of how the basic abductive argument might work in the context of archaeology.

as good as Devitt's basic argument. Second, these two arguments for realism – Devitt's abductive and Hacking's experimental argument – correspond neatly to the two roles that unobservables can play in science. Hacking's argument is available whenever the unobservables are serving as tools for the production of new phenomena, and Devitt's is available whenever the unobservables are serving to unify the phenomena. That is why Devitt's argument is available in the context of historical science. The cumulative case for scientific realism is therefore strongest when these two arguments converge on the same conclusion, which will happen whenever the unobservables are playing both roles simultaneously. We can most confidently assert that we know something about unobservables when they play both roles at once.

One possible response to all this would be to deny that Hacking's argument lends any extra support to realism in the first place. Then the case for both experimental and historical realism would rest entirely with Devitt's basic argument. Someone who takes this view would need to explain what, if anything, is wrong with Hacking's argument, and why that argument lends no extra credence to experimental realism.

3.6 THE CLASSICAL ABDUCTIVE ARGUMENT FOR REALISM: BOYD

Perhaps the most enthusiastically endorsed, most roundly criticized, and most widely discussed argument for scientific realism is the inference to the best explanation of the empirical success of science, which most people trace back to Hilary Putnam's (1978) famous "no miracles" argument and J. J. C. Smart's (1963) "cosmic coincidence" argument. This argument has undergone numerous refinements over the years, and it has been subjected to harsh criticism by Laudan (1981) and others, only to be reformulated and reaffirmed – most recently, for example, by Leplin (1997) and Psillos (1999), whose versions I discuss in chapter 5.

One of the most highly regarded versions of the no miracles argument is that offered by Richard Boyd, who has repeatedly stressed the dialectical interaction between scientific theory and method: The use of heavily theory-dependent methods leads to the development of better theories, which in turn leads to methodological refinements, and so forth. Boyd says that a theory is instrumentally reliable if it yields true predictions about phenomena, and that methods are instrumentally reliable to the extent that scientists who employ them end up accepting instrumentally reliable theories. His abductive argument for realism then proceeds as follows:

> According to the realist, the only scientifically plausible explanation for the reliability of a scientific methodology that is so theory-dependent is a thoroughgoingly realistic explanation: scientific methodology, dictated by currently accepted theories, is reliable at producing further knowledge precisely because, and to the extent that, currently accepted theories are relevantly approximately true. (Boyd 1990, p. 223)

In this passage, Boyd is primarily addressing selective skeptics, such as Van Fraassen (1980), who reject realist epistemology. Boyd's argument is that if we did not already have a good deal of theoretical knowledge of unobservables, then the instrumental reliability of theory-dependent methods would be inexplicable.

Boyd's dialectical version of the no miracles argument is most plausible in the case of experimental science. The more we know about things that are unobservably tiny, the more ingenious will be our experimental apparatus and designs. The better our experimental apparatus and designs, the better able we will be to test our theories about the unobservably tiny things, while being careful to rule out false positives and false negatives. Since the experimental methods are so heavily theory-dependent, their success at producing instrumentally reliable theories would be a miracle if the original theories were not approximately true. It is no coincidence that in one of Boyd's classical statements of his view, the whole argument is couched in terms of experimental methods, experimental design, and the assessment of experimental results (1985, pp. 4–6). For Boyd, the dialectical interaction between theories and experimental methods is what clinches the case for realism. Devitt, who accepts this argument though he does not take it to be the most powerful one in the realist's arsenal, captures the idea nicely when he says that "not only are scientists learning more and more about the world," but that they are also "learning more and more about how to find out about the world" (1991, p. 163). How does this dialectical argument fare in the context of historical science, where the unobservables can only serve as unifiers of the phenomena, and never as tools for the production of new phenomena? I argue that it does not fare quite so well. Perhaps this is no surprise, since it is not clear that Boyd ever intended his argument to be used outside the context of experimental science.

In order to see why the argument does not support historical realism as powerfully as it supports experimental realism, we need to look a bit more closely at its structure. Boyd's dialectic goes as shown in diagram 3.1.

(A)
Theories about unobservably tiny things (e.g. electrons and positrons)

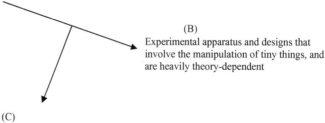

(B)
Experimental apparatus and designs that involve the manipulation of tiny things, and are heavily theory-dependent

(C)
Instrumentally reliable theories about tiny things
(e.g. quarks, fractional electric charges)

Diagram 3.1

The arrows in this diagram represent dependence relations. The argument is that the success of our theory-dependent methods (B) at producing instrumentally reliable theories (C) would be inexplicable if our theories about unobservables (A) were not approximately true, or if experimental realist epistemology were false. The question is whether there are any cases involving a dialectical process that is analogous to the one invoked by the argument for experimental realism (see diagram 3.2).

(A)
Theories about unobservables that existed or occurred in the past.

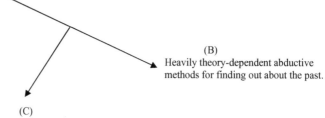

(B)
Heavily theory-dependent abductive methods for finding out about the past.

(C)
Instrumentally reliable theories about unobservables
That existed or occurred in the past.

Diagram 3.2

If this dialectic were present, then it might be possible to generate a convincing Boyd-style argument for historical realism. Consider Padian and Olsen's experimental work, described earlier. Their method involved the testing of biomechanical generalizations by having flightless birds make tracks in different substrates at different gaits and speeds, and then using those generalizations to draw conclusions about ancient therapod dinosaurs. These methods (B) are, to be sure, heavily dependent upon theories about the past. They depend on (A) theories of taphonomy (i.e. theories about the fossilization process), evolutionary theory, biomechanical theories, and so forth. Padian and Olsen's claim that their inferences about ancient therapods are warranted because birds and therapods are closely related phylogenetically involves a number of assumptions about past evolutionary processes. What's more, these theory-dependent methods enable them to identify a smoking gun for the hypothesis (C) that therapods had an upright posture. The trouble is that this last hypothesis does not have a very high degree of instrumental reliability. When conjoined with the biomechanical generalizations, it does imply some accurate predictions about observable therapod tracks, but the set of phenomena predicted by this hypothesis about ancient therapods is coextensive with the set of phenomena that are unified by the hypothesis. Since the tracks are the only relevant traces we have or will ever have, the hypothesis about therapod posture does not lead to any novel predictions in either the temporal sense or the more important epistemic sense – i.e. predictions of phenomena that are different from the phenomena that the hypothesis was originally introduced to explain (see, e.g. Leplin 1997). The whole argument would be much stronger if the resulting theory (C) were more instrumentally reliable – that is, if it implied novel predictions, or even just predictions of phenomena that can be reproduced at will under experimental conditions.

Another important thing to see is that in this case, in contrast to the case described by Hacking, the theory/method dialectic is not likely to continue to unfold in interesting ways. The hypothesis that therapods had erect posture has not helped scientists continue to improve their methods for finding out about the past. One would be hard pressed to identify any current paleontological methods that depend on the theory that therapods had erect posture. While the Boyd-style dialectic is not completely absent from this episode of historical science, neither is it present in all of its richness.

Here then is one initial, though admittedly not very powerful reason for thinking that Boyd's inference to the best explanation of the success of theory-dependent scientific methods will lend less support to historical than to experimental realism. It could turn out that in cases where the unobservables in question existed or occurred in the past, the Boyd-style dialectic between theory and method will only be partial. In order to make this argument stick, it would be necessary to consider a much wider range of examples of theories about past unobservables. This turns out not to be necessary, however, because there is a much more powerful reason for thinking that Boyd's inference to the best explanation lends more support to experimental than to historical realism.

I come now to the main argument of the chapter. Larry Laudan (1981, pp. 133–135) and Arthur Fine (1986a, p. 161) have suggested that Boyd's argumentation is viciously circular.[8] According to these critics, the whole point at issue is whether we may rely on inference to the best explanation to lead us to true conclusions about unobservables. Philosophers who try to defend scientific realism by making an argument to the best explanation are using the very style of argument whose reliability is in question.

If we are considering the dialectical argument for experimental realism, Boyd would seem to have an adequate reply to the above objection: He could point out that IBE involving unobservables is *not* the main thing in question. The theory-dependent methods (B) that figure so prominently in the argument for experimental realism are *not* in the first instance abductive methods; they are, rather, experimental methods, involving the design of experimental apparatus and the use of unobservables to produce new observable effects in the laboratory setting. By setting up the argument as an inference to the best explanation of the success of theory-dependent experimental methods – methods which are not abductive, because they involve using unobservables to produce phenomena, rather than positing unobservables as unifiers of the phenomena – the realist can avoid the charge of vicious circularity. However, the abductive argument from the success of theory-dependent methods will indeed be viciously circular if it is offered as an argument for historical realism. The reason for this is simply that all of our knowledge of past unobservables is abductive to begin with. What we have here is differential vulnerability to the circularity objection.

---

[8] Arthur Fine's own way of stating the circularity objection against the abductive argument for realism makes it look as if he is committed to rejecting abductive inference *tout court*, which I do not think is the case. Fine writes that "Metatheoretic arguments must satisfy more stringent requirements than those placed on the arguments used by the theory in question, for otherwise the significance of reasoning about the theory is simply moot. I think this maxim applies with particular force to the discussion of realism" (1984, p. 23). I take him to be saying that you can't use abduction to justify abduction, which does not by itself imply that we are not justified in using abduction. I am convinced that other aspects of Fine's view actually commit him to the view that non-empirical virtue carries evidential weight. (See, for example, my discussion of the parity principle in ch. 7.1.) Also, Fine writes that "because of its parsimony, I think the minimalist stance represented by NOA marks a revolutionary approach to understanding science," as if its parsimony were a reason for thinking that NOA is correct (1984, p. 101). This is not the sort of thing we would expect a skeptic about abduction to say.

If experimental methods were abductive, then the Boyd-style argument for experimental realism would be just as vulnerable to the charge of vicious circularity as the argument for historical realism. How reasonable is it, then, to claim that experimental methods are not abductive? It would be misleading to suggest that experimenters never reason abductively – for example, that they never infer to the best explanation of experimental results. However, Boyd's dialectical argument for experimental realism is best interpreted as an inference to the best explanation of the success of what experimenters *do*. The experimental methods whose success the realist hopes to explain are methods for building apparatus, manipulating test conditions, and (as Hacking would put it) intervening in nature. It is possible to defend experimental realism in a way that avoids the charge of vicious circularity, so long as one emphasizes the practical nature of the experimental methods.

In order to avoid misunderstanding, it is worth pointing out that realists may be vulnerable to two different complaints about their reliance on abductive reasoning: a general objection and a more specific one. The general objection, as we have seen, is that realists are not entitled to the use of abductive arguments at all. For example, one might worry that if the initial pool of potential explanations of some phenomenon is a *bad lot*, which is to say that all of the potential explanations in the pool are false, then it would be a mistake to conclude that the best explanation in the pool is true or even likelier to be true than not. Unless the realist can address this and other reasons for being skeptical about abduction in general, the abductive arguments for realism will beg the question against the critics. For present purposes, we can leave this general objection to one side, because it has equal force against the abductive arguments for historical and experimental realism.[9] The more specific complaint is that it is viciously circular to attempt to defend realism by arguing that realism affords the best explanation of the success of abductive methods. This more specific objection is the one that Boyd can avoid by arguing that the relevant experimental methods – the methods whose success the realist seeks to explain – are not themselves abductive.

Here, then, is the main argument:

P1. Things that are unobservable because they existed or occurred in the past can only play the unifying role, and can never be used as tools for the production of new phenomena.

---

[9] For one version of this argument from the bad lot, see van Fraassen (1989, pp. 149–150). For a reply to this argument, see Lipton (1993).

C1. The methods used to learn about past unobservables will therefore typically, if not always, be abductive; that is, they will typically involve inferences to the best explanatory unification.

C2. Therefore, what is at stake in the debate about historical realism (and in particular, historical realist epistemology) is the reliability of abductive inferences to conclusions about past unobservables.

P2. The classical inference to the best explanation of the success of historical science is itself an abductive inference to a conclusion about past unobservables.

C3. Therefore, the classical abductive argument for historical realism is viciously circular.

The dialectical defense of experimental realism is not circular in this sense.

Some philosophers might admit that the Boyd-style argument for historical realism is circular, while denying that it is viciously so. For example, it could be that the argument is rule-circular but not premise-circular, whereas only the latter kind of circularity is vicious. Stathis Psillos (1999) takes this line, arguing that scientific realism should be seen as part of a larger philosophical package that includes epistemological externalism. Externalism implies that rule-circular arguments are not in general vicious. Or it could be that abduction is a non-transitive form of inferential justification. John Post (1996) makes this move, arguing that there can be closed loops of non-transitive inferential justification that are not necessarily vicious. However, if the only difference between two arguments A and A^*, is that A is circular while A^* is not, then presumably one would say that A^* is the better argument, even if the circularity exhibited by A is not vicious. Therefore, even if the circularity of the Boyd-style argument for historical realism is not a sufficient reason for rejecting that argument altogether, it is still a reason for thinking that it is weaker than the corresponding argument for experimental realism.

The question of when, exactly, a circular argument is viciously circular, or when exactly an argument begs the question, is a vexed one. It is certainly not a question that I can hope to settle here. Fortunately, however, we do not need to settle it at all, at least not for present purposes. Since the Boyd-style defense of experimental realism, when it is formulated in the right way, is not circular at all, it is (contrary to Laudan and Fine) not vulnerable to the charge of vicious circularity in the first place. My claim is that the abductive argument for historical realism, but not the abductive argument for experimental realism, is vulnerable to the charge of vicious circularity. This alone is reason for thinking that the one argument is less good than the other.

Now let's put the two arguments of this chapter together:

P1. We can be most confident that we know something about unobservables when those unobservables play both the unifying and the producing roles. But since past unobservables can never play both roles, our claims to know something about the past are less secure than our claims to know something about the tiny.

P2. The Boyd-style inference to the best explanation of the success of science is vulnerable to the charge of vicious circularity when used to support our claim to know something about prehistory, but it can avoid this charge in the context of experimental science.

C. Therefore, if the traditional arguments for realism are any good at all, they give less support to minimal epistemic realism about the past than to minimal epistemic realism about the tiny.

This last conclusion (C) is what I mean by *historical hypo-realism*.

3.7 MCMULLIN ON FERTILITY AND METAPHOR IN SCIENCE

Ernan McMullin (1984) is a rare realist philosopher who draws upon examples from historical science in developing an argument for scientific realism. For this reason, his argument deserves special attention.

To begin with, McMullin focuses on cases in which the history of science seems to exhibit a "progressive discovery of *structure*" (1984, p. 26, his italics). He suggests that scientists typically explain phenomena by coming up with "models of the hidden structure of the entities being studied." This hidden structure is then supposed to provide a causal explanation of the phenomena. So far, this sounds a lot like Devitt's basic abductive argument for realism, with a slight shift of emphasis from unobservable things and events to unobservable structures.

The classic realist move is to insist that there is some fact about science that is best explained by supposing that we can and do know something about unobservable things, events, and/or structures. McMullin argues that anyone who rejects realism will have a difficult time explaining the *fertility* of our models of the hidden structure of the world. He argues that "such fertility finds its best explanation in a broadly realist account of science" (1984, p. 34).

To illustrate what he means by fertility, McMullin recounts the story of the transition from Alfred Wegener's theory of continental drift to the modern theory of plate tectonics. Wegener first introduced his theory in order to explain certain puzzling phenomena, such as the fact that the coastlines of South America and Africa seem to fit together like puzzle pieces. From the very beginning, however, the theory of continental drift confronted some serious anomalies. In order to move, the continents would have to cut through the ocean floor, and how was that supposed to work? Still, although no one today believes in the original version of the theory of continental drift that Wegener put forward, the theory turned out to be extremely fertile. McMullin argues that the model, which involved continents sliding around the earth's surface, had a great deal of "metaphoric power." The idea was suggestive, and during the 1960s, geologists developed the theory of plate tectonics. Though inspired by Wegener's theory of continental drift, the theory of plate tectonics handled the anomalies that made it so difficult for people to take continental drift seriously.

> The original theoretical entity, a floating continent, did not logically entail the plates of the new model. But in the context of anomalies and new evidence, it did suggest them. (1984, p. 32)

McMullin also emphasizes the continuity between the theory of continental drift and its successor, the theory of plate tectonics:

> The important thing to note is that there are structural continuities from one stage to the next, even though there are also important structural modifications. What provides the continuity is the underlying metaphor of moving continents that had been in contact a long time ago and had very gradually developed over the course of time. One feature of the original theory, that the continents are the units, is eventually dropped; other features, such as what happens when the floating plates collide, are thought through and made specific in ways that allow a whole mass of new data to fall into place. (1984, p. 33)

McMullin is surely right to call attention to the fertility of the theory of continental drift. He is also right to point out that this fertility has less to do with the logical consequences of that theory, and more do to with the fact that certain metaphorical extensions of the theory later turned out to be wildly empirically successful. Next, he asks what explains the fertility of a theory such as Wegner's theory of continental drift? What explains the fact that it turned out to be so profitable for scientists to explore the metaphorical extensions of that theory? The best explanation, according to McMullin, is just that there was something right about continental drift. The structure posited by the theory of continental drift – the moving continents – is something like the real structure of our planet.

McMullin's argument for realism, like Boyd's, has a dialectical flavor. But whereas Boyd focuses on the dialectical relationship between theories and methods in science, and on the way in which theories give rise to new methods, McMullin focuses on one way in which theories give rise to new theories. Boyd looks at theory-dependent methods and argues that it is hard to believe that any theory that is not true or approximately true could give rise to such successful methods. Although he does not put it in these terms, McMullin invites us to think of plate tectonics as a theory-dependent theory. We can then see him as arguing that unless there were something right about the theory of continental drift, it is hard to see how that progenitor theory could have given rise, via metaphorical extension, to such a successful offspring. We can thus see how McMullin tries to extend the traditional realist line of argument in a new way.

There are, however, two problems with McMullin's approach. The first problem is that we can already give a straightforward Devitt-style abductive argument for realism about tectonic plates. All we would need to do is to list all the otherwise puzzling phenomena that are best explained by supposing that tectonic plates move gradually. And the list would be a long one: earthquakes, mountain ranges, undersea trenches, patterns of volcanic activity, volcanic island chains, seafloor spreading, and much else. In the case of plate tectonics, this basic abductive argument is already very strong, and we add nothing to it by pointing out that the theory which preceded plate tectonics had the virtue of fertility. And this leads me to the second problem for McMullin's approach, which is that he gives us an argument for being realists *about the wrong theory*. What we want is an argument for taking the realist view, say, of plate tectonics – that is, an argument for thinking that the theory of plate tectonics is true or approximately true. McMullin suggests that the fertility of the earlier theory – continental drift – is what cries out most loudly for realist explanation. But this approach yields an argument for taking the realist view of a theory that has been superseded.

Perhaps McMullin could respond to this worry by arguing that the theory of plate tectonics is even more fertile than its predecessor, the theory of continental drift, so that the case for the realist interpretation of plate tectonics is even stronger. The problem with this response is that we are not yet in a position to assess the fertility of plate tectonics. To do that, we would need to look for anomalies that crop up, and see whether future theories which are metaphorically suggested by plate tectonics succeed in handling those anomalies. Only then could we mount a McMullin-style argument for a realist interpretation of plate tectonics. Notice, however, that by that point, plate tectonics would itself have become the predecessor theory. Although the argument for realism as the best explanation of fertility might well go through, it would be misdirected: Realism is supposed to be a view about current theories, not about predecessor theories.

McMullin actually distinguishes two kinds of fertility. In addition to the kind of theoretical fertility that I have been discussing here, there is another kind of fertility that has do to with novel predictions. This second kind of fertility is so important to the case for scientific realism that it will take up chapter 5. For now, though, I hope to have shown that manipulation matters. The fact that we cannot manipulate the past weakens the abductive arguments for realism about prehistory.

4

Paleontology's chimeras

Scientists who want to reconstruct the past often use presently existing things as models for things that no longer exist. For example, since we cannot observe the social lives of our Pleistocene ancestors, one suggestion is to treat existing hunting and gathering societies as models. Since we cannot observe dinosaurs, we might use living birds and mammals as biomechanical models. Up to now I have been arguing that historical science finds itself at a relative disadvantage, because (a) we cannot manipulate the past, and (b) historical processes destroy information about the past. But maybe these disadvantages are counterbalanced by another special advantage: the ready availability of observable analogues for prehistoric entities and events.

4.1 THE ANALOGUE ASYMMETRY

Although he does not use this terminology, Christián Carman (2005) calls attention to another potentially relevant difference between the past and the tiny. Picking up on an argument from Rom Harré (1986; 1996), Carman suggests that while there are plenty of observable analogues for past things, events and processes, there are few if any observable analogues for microphysical things, events, and processes. The particles and processes described by quantum theory are so wildly different from any of the middle-sized dry goods that we encounter in ordinary life, that scientists cannot safely use observable things and events as a guide to the ontology of the microphysical universe. On the other hand, the plants, animals, and natural processes that we can observe on Earth today are not so radically different from past living things and processes. Although scientists obviously need to proceed with caution when comparing the

85

present to the past, still it seems reasonable to suppose that the observable analogues give historical scientists more guidance than they give to scientists who want to study the microphysical structure of the universe.

As with the asymmetries discussed earlier – the asymmetry of manipulability and the role asymmetry of background theories – we can pose several questions about this one: First, is it a real asymmetry? Second, if this is a real asymmetry, what explains it? Third, what if any interesting epistemological consequences does it have? Carman argues, in effect, that the asymmetry of observable analogues confers an epistemic advantage on historical science.

Carman himself states this argument in connection with the subject of underdetermination:

> Harré claims that scientists, when proposing models in order to eliminate underdetermination, use the strategy of drawing inspiration from well known and accepted ontologies of entities that we have already been able to observe. Infinitely different models could explain the same phenomena, but very few of them are ontologically plausible. The inspiration in well known ontologies grants the models certain plausibility, since they resemble things, processes, events, and so on that really exist (or existed) . . . To put it in a nutshell, there is a higher risk in putting forward a possible ontology for a proton than in doing the same for a dinosaur, since we have observed entities that are similar to the dinosaurs but not to the protons. (Carman 2005, p. 173)

This interesting line of argument spells trouble for much of what I have said in previous chapters. For example, in chapter 2 I claimed that local underdetermination will be a more pervasive problem in historical than in experimental science, but Carman has given us a reason to think otherwise. In domains where we have a supply of observable analogues, we can rule out many hypotheses and theories as ontologically implausible. This means that underdetermination might be a bigger problem in experimental science than in historical science.

One initial response to this problem is that the local underdetermination problems that figure so prominently in chapter 2 arise in cases where we have reason to think that we will probably never have the evidence required to discriminate between two hypotheses, both of which have about the same degree of ontological plausibility. The fact that historical scientists can rule out many hypotheses on grounds of ontological implausibility therefore does not alter the conclusion that local underdetermination will be especially pervasive in historical science. This initial response is not entirely adequate, though. For the ability to rule out some ontologies as implausible could still give historical science an epistemic advantage vis-à-vis experimental science. That is, if local underdetermination is a serious problem even when we can rule out many hypotheses as being ontologically implausible, it could be an even worse problem in other areas of science where the lack of observable analogues for unobservable entities and processes makes this initial filtering impossible.

A better response is to point out that the relative abundance of observable analogues (usually artifacts and living organisms) is something of a mixed blessing in paleontology. Analogies between ancient organisms and living organisms have often misled scientists in the past. I will use several case studies from the history of paleontology to illustrate the ways in which analogies, though indispensable, have led scientists into persisting errors. Because observable analogues have the potential to mislead, the asymmetry of observable analogues does not give historical science any special epistemic advantages vis-à-vis experimental science.

Another possible line of response to the above argument, though not one that I will pursue here, would be to deny that the asymmetry of observable analogues is a real asymmetry in the first place. Scientists frequently do use analogies to aid their theorizing about the microstructure of the universe: DNA molecules resemble spiral staircases; atoms resemble the solar system; the molecules of a gas move and interact like tiny billiard balls; the basic constituents of matter vibrate like the strings of a musical instrument, and so on. For the sake of argument, however, I will assume that there really is an analogue asymmetry – that is, that while we have an abundance of observable analogues for past things and events, we have relatively few observable analogues for unobservably tiny things and events. The issue I want to explore is whether this analogue asymmetry would confer any epistemic advantage on historical science.

4.2 MISLEADING OBSERVABLE ANALOGUES IN PALEONTOLOGY[1]

For several decades during the twentieth century, scientists remained quite convinced that the duckbilled dinosaurs (the lambeosaurine

---

[1] Although I focus exclusively on paleontology in this chapter, I think the lessons about the ways in which observable analogues can mislead will generalize to other sciences as well. For a related discussion of the role of analogical reasoning in archaeology, see Salmon (1982, chapter 4). Of particular interest is her assessment of the role of analogy in ethnoarchaeology, or the ethnographic study of present cultures for the purpose of drawing archaeological inferences (1982, pp. 74ff.).

hadrosaurs) lived an aquatic or semi-aquatic lifestyle. The duckbills were frequently pictured foraging for food along the bottoms of shallow lakes and rivers, or retreating to the water in order to avoid predators. This interpretation was supported by fossilized skin impressions – including one famous dinosaur "mummy" – that clearly indicated that the animals had webbed feet, just like ducks. The obvious conclusion, based upon the analogy between the unobservable dinosaur and the observable living organism, later turned out to be badly mistaken (see Turner 2000, for the full story). The flaps of skin that look like webs in the fossilized impression were probably more like the pads on the bottoms of camels' feet – adaptations for overland travel, rather than for paddling through the water (Bakker 1986, chapter 7). The cranial crests of these organisms contained elaborate hollow sinus passages and cavities. During the early to mid-twentieth century, many scientists assumed that the cranial crests must have functioned as snorkels or as air tanks that would have enabled the animals to remain submerged while foraging – another analogy that later proved to be misleading. The crests have since been reinterpreted as resonating devices and/or display structures that might have helped individuals recognize conspecifics in mixed species herds (Hopson 1975; Weishampel 1981, 1997).

The story of *Anomalocaris* (as told by Gould 1989; Collins 1996) affords another example of the multifarious ways in which presently observable organisms and artifacts have duped paleontologists. In 1886, the Geological Survey of Canada dispatched Richard McConnell to map the geology along the route of the recently finished Canadian Pacific Railway. In the small town of Field, British Columbia, he heard rumors about some "stone bugs" that had been found in the nearby mountains. When McConnell went to investigate, he discovered, in addition to numerous trilobite fossils, unusual fossilized body parts of some previously unknown marine organisms. Joseph Whiteaves, the chief paleontologist for the Geological Survey of Canada, published the first description of these strange fossils (Whiteaves 1892). Since the closest observable analogues Whiteaves could find for the strange fossilized parts were the tails of living shrimp, he classified this Cambrian organism as a crustacean and named the creature *Anomalocaris canadensis* (meaning "anomalous Canadian shrimp"). Whiteaves noted that the fossilized tails are not exactly like the tails of living shrimp. The appendages sticking down from them are unjointed, whereas most crustaceans have jointed appendages. Nevertheless, he thought the resemblance strong enough to warrant their classification as crustaceans. Almost a century passed before Whittington and

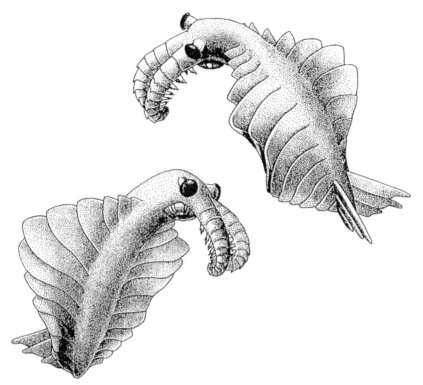

Figure 4.1 Reconstruction of *Anomalocaris Canadensis*. Joseph Whiteaves originally mistook a fossilized impression of one of the feeding appendages for the
tail of a shrimp.

Briggs (1985) revealed the error by showing that what Whiteaves had
identified as a shrimp tail was really the feeding appendage of the considerably larger marine predator shown in figure 4.1. The history of paleontology is replete with such examples of chimeras – creatures, cobbled
together out of the parts of other organisms, that scientists imagined (and
with good reason) to have existed long ago, but which never did really
exist.

Not all scientific errors are interesting. Some of them are mere goofs,
and these are usually corrected in short order by the scientific community.
One example of a goof from the history of paleontology is the incident
that sparked the great feud between Edward Drinker Cope and Othniel
Charles Marsh. In 1868, Cope had received a specimen of a long-necked
plesiosaur from one of his diggers in Kansas. Cope named the animal
Elasmosaurus platyrus and set about reconstructing it for the Academy

of Natural Sciences in Philadelphia. Up to that point, Cope's correspondence with Marsh, who was based in New Haven, Connecticut, had been entirely cordial. In 1869, Marsh visited Cope in Philadelphia to view the reconstruction of *Elasmosaurus*, and pointed out to the latter that he had mounted the animal's vertebral column backwards. Cope had mounted the animal's skull on the tip of its tail. Cope did not take the correction very well (Jaffe 2000, pp. 12–14). Cope's error is an example of what might be termed a *shallow mistake*. Even Cope himself recognized the mistake for what it was as soon as it was pointed out to him, and he revised his reconstruction shortly thereafter. Shallow mistakes like this one hold less interest for philosophers of science. More important are the *deep mistakes*, or false conclusions which are believed by large numbers of scientists, and for longer periods of time.

There are many other fascinating cases in which scientists were misled by observable analogues for prehistoric creatures. For instance, the great nineteenth-century paleontologist Richard Owen once mounted a thumb spike of the dinosaur, *Iguanadon*, on the tip of its nose after noting the resemblance between the teeth of the dinosaur and those of a small horned iguana (Cadbury 2000, p. 155). I will describe just one more case in detail. In 1899, an expedition from the new Carnegie Museum in Pittsburgh found the remains of an 87-foot long sauropod dinosaur in Jurassic rocks at Sheep Creek, Wyoming (Bakker 1986, chapter 10; see also Parsons 2001). John Bell Hatcher, the leader of the expedition, named it *Diplodocus carnegiei*, after his patron Andrew Carnegie, and shipped it back to Pittsburgh. Carnegie, who was eager to show off his new dinosaur, had complete plaster casts of the skeleton made and delivered to King Edward VII, Kaiser Wilhelm, and Georges Clemenceau. Casts of *Diplodocus carnegiei* eventually ended up in Bologna, St. Petersburg, Vienna, Mexico City, and La Plata, Argentina. This meant that scientists all over the world had the chance to study the same dinosaur skeleton. Not surprisingly, they soon arrived at different conclusions.

The original at the Carnegie Museum in Pittsburgh was reconstructed by John Bell Hatcher, and, after Hatcher's death in 1904, by William J. Holland. Hatcher and Holland, building upon the earlier work of Marsh and Cope, proceeded on the assumption that the sauropod dinosaurs' legs were like columns. When the giant animals walked, the legs swung directly below the torso, as in living elephants and rhinoceroses. A number of paleontologists in Europe and a few in the United States rejected this elephant model in favor of what can only be described as a "lizard model." The German scientist Gustav Tornier argued that since dinosaurs were reptiles, *Diplodocus* should be reconstructed with its legs sprawling out to the sides, like the legs of a lizard or an alligator. At the Senckenberg Museum in Frankfurt, one copy of a *Diplodocus* (a member of a species that had been discovered somewhat earlier by Cope) had already been built according to the elephant model. It was now dismantled and put back together again so as to fit Tornier's lizard model. Both the Americans and the Germans were using analogical reasoning to guide their reconstruction of the extinct organism, but they disagreed about which living organism provides the best analogue for *Diplodocus*.

The American and the German scientists poked fun at one another's reconstructions of *Diplodocus*, with Holland at one point referring to Tornier's lizard model as "a skeletal monstrosity." He joked that given the length of the animal's legs and the depth of its torso, in order to get from place to place, the German *Diplodocus* would have had to find giant ruts in the ground.

Most German scientists gave up on the lizard model once sauropod bones began to arrive in Berlin from the newly discovered bone quarries at Tendaguru Hill in what is now Tanzania (but at the time was German East Africa). A German expedition led by Werner Janensch quarried Tendaguru Hill for four years, from 1909 to 1912. During that time, 250 tons of bones were carried by local porters from Tendaguru to the town of Lindi, the nearest seaport, from whence they were shipped back to Berlin. The German expedition's biggest find was *Brachiosaurus*, a sauropod only distantly related to *Diplodocus*, and with a rather different body architecture, and one of the possible wide-gauge trackmakers mentioned in chapter 1. But the discoveries in Africa led continental paleontologists to rethink their reconstruction of the sauropods.

In the end, though, the lizard model was laid to rest by the discovery of footprints. In the late 1930s, Roland T. Bird, who got his start working under the famous fossil hunter Barnum Brown for the American Museum of Natural History, found the first confirmed sauropod trackways. It was, and still is impossible to say exactly which species made which set of tracks. Bird's trips to the Paluxey River in Texas were paid for, not coincidentally, by the Sinclair Oil Company (see Farlow and Lockley 1989). Even the widest-gauge trackways that have been found, however, are far too narrow to have been made by sprawling sauropods of the sort imagined by Tornier. Bird's discoveries in Texas, and others made since then, strongly disconfirm the lizard model. Indeed, the lizard model is an excellent example of an architectural hypothesis that failed a risky empirical test, of Popperian falsification in historical science.

4.3 EXPLAINING PAST SCIENTIFIC MISTAKES

Many scientific realists, as we have seen, take the view that one of the main functions of philosophy of science is to provide an explanation of the empirical success of science. This concern reflects the fact that nowadays most philosophers of science embrace some form of philosophical naturalism (Rosenberg 1996). Naturalism is notoriously difficult to characterize with much precision, but many naturalists share the methodological view that philosophy of science should be an empirical inquiry, no different in principle from empirical science itself. Philosophers of science begin with a puzzling fact about the natural world – namely, that human scientists have achieved astounding predictive successes – and they try to generate and test potential explanations of this fact. This activity does not differ much from that of scientists, who also generate and test potential explanations of puzzling phenomena in nature.

But if the empirical successes of science require some sort of explanation, what about the failings? What about cases, such as those described above, in which some of the very best scientists of the day persisted in believing false hypotheses, and for a considerable length of time? In these cases from the history of paleontology, the observable analogues available to scientists – the snorkels, the living ducks, shrimp, lizards, and so on – help explain the mistakes. In each case, scientists drew false conclusions precisely because they judged that some long extinct creature must have been rather similar to some living organism, or that some part of an extinct creature must have resembled some familiar artifact.

Imagine that an ethnographer visits a community of hunters and gatherers who have never before seen an automobile. The ethnographer presents the community with a gift: a box of battered, disassembled car parts, including a spark plug, a steering wheel, a hub cap, and so on. But the ethnographer also refuses to explain where the parts came from. The recipients of this gift conclude that the hub cap must be a shield, or a piece of headgear, or a serving platter, and they persist in this mistake. If we wanted to explain why the people in this community held a mistaken belief about the hub cap, we would point out that they had no choice but to generate hypotheses by considering a range of available analogues (shields, platters, etc.). Another crucial part of the explanation would simply be that they do not have access to certain kinds of empirical evidence – i.e. that they have never seen an assembled car.

Now let's modify the thought experiment in order to highlight the ways in which missing evidence and observable analogues can work together to mislead. Suppose that the ethnographer, for whatever reason, wants to find out whether the members of the hunting and gathering society would be able to reconstruct a car from spare parts. She visits three different communities. The first group receives as a gift nothing but a hub cap, and is then faced with the challenge of reverse engineering this particular car part. The second group receives a shipping crate full of car parts, but the set of parts is incomplete. For example, all the tires have been removed, as have other crucial parts, such as, say, the chassis. Finally, the third community receives a complete set of car parts. Clearly, this is an unfair contest. Perhaps no one will be able to reconstruct the car with much accuracy, but the members of the third group – the ones who received a complete set of parts – are the only ones who stand much of a chance.

Next, suppose that the ethnographer gives a complete set of car parts to each of the three communities, but that this time she also provides each group with an analogue that will help guide their reconstruction of the car. The first group receives a lawn mower, the second receives a horse-drawn buggy, and the third receives a car, though one of a different make and model than the disassembled car. Once again, the contest is obviously unfair. There are some important similarities between a lawn mower and an automobile; both contain internal combustion engines, for instance. Likewise, there are some important structural similarities between a horse-drawn buggy and a car; perhaps both contain bench seats. Yet these analogues are also liable to mislead the people charged with the task of rebuilding the car, and to mislead them in all sorts of unpredictable ways. The lesson to be drawn from these thought experiments is that there are two main sources of mistakes in paleontological reconstruction: evidential incompleteness and inadequate observable analogues.

Now we can run the following argument, which is a relative of Larry Laudan's (1981) famous pessimistic induction from the history of science. The argument is based (ironically) on an analogy between past and present scientific episodes:

P1. In a range of past cases, paleontologists drew mistaken conclusions about ancient organisms. These mistakes were made by the very best scientists and took a long time to correct.

P2. Part of the best explanation of such mistakes is that the scientists generated their hypotheses about the ancient organisms analogically – that is, by supposing that the ancient organisms must have resembled some observable organisms or artifacts.

P3.   Current paleontological reconstructions of ancient life are no different from past ones, insofar as scientists must still generate hypotheses by analogy.

C.   We should therefore be rather skeptical about the best current reconstructions of ancient organisms.

This argument is anything but decisive, and I do not wish to ask it to carry too much of a load. The basic idea is that one factor which contributes to the explanation of mistakes in past cases is also at work in present cases. One obvious problem with it is that there are plenty of relevant differences between the past and present reconstructions of ancient organisms. For example, there are just more fossils available to researchers today than there were in the past, and specimens of higher quality. Indeed, several of the mistakes described above were finally corrected by the discovery of better fossilized remains. The important thing to see about this argument is that it turns Carman's earlier argument on its head. Carman suggests that the availability of observable analogues confers an epistemic advantage that makes historical science better off, in at least one respect, than experimental science. The observable analogues are supposed to be advantageous because they constrain the range of plausible ontologies that scientists may posit in order to explain and/or predict the phenomena. However, the version of the pessimistic induction that I have just constructed shows that the availability of observable analogues is not wholly advantageous from an epistemic point of view. The observable analogues are a mixed blessing at best, even granting Carman's claim that their relative abundance makes underdetermination less of a problem in historical science than in experimental science. Although paleontological reconstruction would scarcely be possible without them, they have also contributed to many past mistakes. Since observable analogues have been known to mislead, having more of them is not necessarily a good thing, from an epistemic point of view.

Thus far, I have argued that the analogue asymmetry (assuming that it is a real asymmetry) would not give any epistemic advantage to historical science or strengthen the case for historical realism. But this leaves open some interesting and relevant questions about the analogue asymmetry. Would this asymmetry place historical science at any epistemic *dis*advantage relative to experimental science (as does the role asymmetry of background theories)? Would it weaken the case for historical realism relative to that of experimental realism (as does the asymmetry of manipulability)? Not necessarily. In fact, I am not sure that very much of interest would follow from the analogue asymmetry, because the important thing is not how many observable analogues we have for the unobservables in a given domain, but whether the observable analogues resemble those unobservables in the right sorts of ways. I propose to drive this point home with two final manipulations of the thought experiment involving the ethnographer and the hunter/gatherer communities.

Suppose, this time, that the ethnographer visits two different communities, and presents each community with the same gift: a set of shipping crates full of damaged, disassembled car parts. However, the ethnographer gives the two communities different sets of observable analogs for inspiration. Community A receives a model car. Community B receives a lawnmower, a model airplane, and a horse-drawn buggy. Here there is an analogue asymmetry between the two communities, because community A has fewer observable analogues to work with than community B does. But this asymmetry lends no particular advantage to the members of community B, because their observable analogues are likelier to lead them to mistaken conclusions about the target item – i.e. about the car. This manipulation of the thought experiment reinforces the point I made earlier by introducing a version of the pessimistic induction from the history of science.

Suppose, next, that community C and community D receive the same gifts as the first two communities. However, community C receives a model car, a lawnmower, and a horse-drawn buggy for inspiration, while community D receives only a model airplane. In this case, we can see that community C would not be at any particular disadvantage relative to community D in virtue of having more observable analogues to work with. On the contrary, the members of community C would have better prospects for coming up with a reasonably accurate reconstruction of the car because their set of observable analogues contains something that resembles the target item in the right sort of way. This remains true even though community C has potentially more misleading analogues to work with than does community D. I conclude, then, that the analogue asymmetry – if it is an asymmetry at all – does not by itself have any interesting epistemic consequences. If anything, this finding should serve to underscore the specialness of the asymmetry of manipulability and the role asymmetry of background theories. Asymmetries between the past and the tiny need not have interesting epistemological consequences.

4.4 THE ANALOGUE ASYMMETRY AND THE
PESSIMISTIC INDUCTION

Part of my strategy for showing that the analogue asymmetry has no interesting epistemological consequences involved the construction of a version of the pessimistic induction from the history of paleontological mistakes. The pessimistic induction (or pessimistic meta-induction, as it is sometimes called) also figures prominently in the scientific realism debate as one of the main counterarguments to the realists' inference to the best explanation of the success of science. For that reason alone, the pessimistic induction deserves a closer look. Does the pessimistic meta-induction have as much force in the context of historical science as it has in the context of experimental science? Do the asymmetry of manipulability or the role asymmetry of background theories have any effect on the strength of the pessimistic induction? Consider the following three hypotheses:

a.   The pessimistic induction has just as much force against historical realism as it has against experimental realism.
b.   The pessimistic induction has more force against experimental realism than it has against historical realism.
c.   The pessimistic induction has more force against historical realism than it has against experimental realism.

Given everything I have said so far, one might expect me to argue at this point that hypothesis (c) is true. However, a good case can be made that (a) is in fact the true hypothesis, and that the asymmetry of manipulability and the role asymmetry of background theories have no effect on the strength of the pessimistic meta-induction. In other words, those two asymmetries have consequences for some of the realism arguments, but not others.

In order to see why this is so, it will help to begin by clarifying the structure of the argument. Although the basic argument has been around for ages – versions can be found in the writings of the ancient skeptic, Sextus – the argument entered the scientific realism debate thanks to Larry Laudan (1981). Laudan set out to attack the realist views of philosophers such as Hilary Putnam, W. H. Newton-Smith, and Richard Boyd, among others, and he did so by producing a now famous list of failed scientific theories:

> the crystalline spheres of ancient and medieval astronomy;
> the humoral theory of medicine;

the effluvial theory of static electricity;
"catastrophist" geology, with its commitment to a universal (Noachian) deluge;
the phlogiston theory of chemistry;
the caloric theory of heat;
the vibratory theory of heat;
the vital force theories of physiology;
the electromagnetic ether;
the optical ether;
the theory of circular inertia;
theories of spontaneous generation. (Laudan 1981, p. 122)

Laudan, much to his credit, did mention at least one historical theory – namely, that of the great flood, which during the nineteenth century was invoked to explain all sorts of phenomena that we now attribute to glaciation, among other causes. There are other failed historical theories that we could add to Laudan's list. Take, for example, contractionist geology, one of the predecessors of the modern theories of continental drift and plate tectonics. According to the contractionist theory, the earth started out much hotter and bigger than it is today. As it gradually cooled, a hard crust was formed around the outside. As everyone knows, things tend to contract when they cool; as the interior of the earth continued to cool and contract, hollow spaces formed beneath the crust. At various points in time, parts of the crust collapsed into these hollow spaces, thus forming the mountains and basins that we see on the surface of the earth today. Another failed historical theory that we could add to Laudan's list is the preformationist theory. According to one version of that theory, all human beings are descended from Adam and Eve. When God created Adam, he also created unobservably tiny versions of Adam's children (i.e. homonculi), and placed them inside Adam's reproductive organs. God placed even tinier versions of their children inside of their tiny reproductive organs, and so on and on. According to this theory, God created all human beings who would ever live at the time he created Adam and Eve; all human beings were preformed, as it were, and the process of development was taken to be a process of growth.

All the theories in Laudan's list have two remarkable features in common. First, they were all badly mistaken. They all incorporated false claims about unobservable entities, events and processes, and those false claims were central, rather than merely incidental to the theories in question. The second important feature that all those theories have in common is that in their day, they enjoyed a high degree of empirical success. Scientists used them to generate lots of accurate predictions, and to explain diverse natural phenomena. Although scientists now know that those theories were mistaken and have since replaced them with better ones, in the past it would not have been irrational for the scientific community to believe many of their respective claims about unobservables. For example, given the evidence available during the late eighteenth century, and given the lack of any serious competitor to the phlogiston theory, it was by no means irrational for leading scientists to believe that phlogiston is released during combustion. At the time, phlogiston theory provided a comprehensive explanation of combustion and rusting, and metallurgists relied on it. This second feature of the theories on Laudan's list – namely, that they were all highly successful in their day – is what makes the list such a challenge for scientific realists.

There are at least three different ways of parlaying Laudan's list into an argument against scientific realism. The first and perhaps most intuitive strategy is to draw an inductive generalization from past cases: Since there are so many past cases in which the central terms of successful theories failed to refer, we have good reason to doubt whether the central terms of our most successful current theories refer to anything in the world. A second strategy is to construe the pessimistic meta-induction as an argument by analogy. Here the first step is to identify some relevant similarity between past episodes and our present epistemic situation. For example, it could be that the sources of error in the past cases are still in play. Since the past and present cases are relevantly similar in this respect, we may conclude that the present theories make false claims about unobservables, just like the past ones. The third strategy – also pursued by Laudan (1981) – is to use the examples on the above list to undermine the realists' inference to the best explanation of the empirical success of science. All of the theories on Laudan's list enjoyed a high degree of empirical success. What could possibly explain that success? Notice that in these cases, we cannot appeal to the truth or approximate truth of the theories in order to explain their empirical success, for the simple reason that the theories were not true or approximately true. That means that we need to cast about for some other explanation of the empirical success of the theories on Laudan's list, and one that makes no appeal to truth or approximate truth. But any explanation we come up with could also be used to explain the success of our best current theories, without mentioning the truth or approximate truth of those theories, and this would undercut the abductive argument for realism. Notice that the main difference between the second and third strategies is that one focuses on explaining scientific failures, while the other focuses on explaining empirical success.

No matter which of these three strategies one adopts, the asymmetry of manipulability will have no bearing on the strength of the pessimistic meta-induction. In order to see why, consider a slightly expanded version of Laudan's list that includes more historical theories such as contractionist geology and the theory of the Noachian deluge. Some of the theories on the list will make claims about manipulable unobservables; others will make claims about unmanipulable unobservables. But this difference will not affect the two crucial facts about the theories on the list that enable the pessimistic meta-induction: (1) the fact that they were all empirically successful, and (2) the fact that their central claims about unobservables turned out to be false. This same point applies to the role asymmetry of background theories. Even if it is true that the background theories of historical science typically tell us how natural processes destroy empirical evidence, while the background theories of experimental science tell us how to create new evidence, this will not change the fact that the theories on Laudan's list all have the two features just mentioned.

In saying that the asymmetry of manipulability does not affect the strength of the pessimistic induction, I mean to leave it open just how strong that argument is. Scientific realists have developed some interesting replies to that argument, and it is worth mentioning three here. First, some have shifted the focus from empirical success more generally to novel predictive success, arguing that it is far more difficult to produce historical examples of false theories that enjoyed novel success in their day. I discuss this line of argument in the next chapter. A second move is to point out that proponents of the pessimistic induction tend to treat confirming evidence and disconfirming evidence differently.[2] They take disconfirming evidence very seriously. After all, the theories on Laudan's list have all, presumably, been disconfirmed. However, proponents of the pessimistic induction seem to give less weight to the confirming evidence that counts in favor of our current theories. Why give more weight to disconfirming than to confirming evidence? Third, one could point out that there are relevant differences between the epistemic situations of earlier scientists and scientists' epistemic situation today. One could argue that scientific methods have improved, and that scientists have gotten better at learning about unobservables. This is certainly true of historical science. Not only do geologists and paleobiologists today have a much

---

[2] I thank an anonymous reviewer for calling my attention to this argument.

larger evidence base than their nineteenth-century predecessors, but they also have access to kinds of evidence that natural historians scarcely could have dreamed of. Take, for example, the drilling technology that enables scientists to extract ice cores and study the chemical composition of bits of air that have been trapped in the ice for many thousands of years. Scientists can sometimes use technology to compensate for nature's destruction of the evidence. Nor are methodological improvements always technological. Today, biologists use cladistic methods to reconstruct evolutionary relationships, and these powerful methods were not available a century ago. These important methodological improvements generate disanalogies between the epistemic situations of past and present scientists, and such disanalogies weaken the pessimistic induction.

At first glance, one might think that the pessimistic meta-induction becomes self-defeating when it is used to target historical realism. The problem is that the premises of the argument all involve claims about the unobservable past – that is, about past episodes in the history of science. If the argument itself is supposed to raise doubts about claims about the unobservable past, then the argument (if it is a good one) would raise doubts about its own premises! Any argument which, if it were good, would give us reason to disbelieve its own premises is a self-defeating argument. However, this line of argument can be blocked by means of the distinction between history and prehistory. We might say that the point of the pessimistic induction against historical realism is merely to raise doubts about claims that scientists make about the unobservable prehistorical past. To do that, the argument relies on claims about the history of science. But these claims about the history of science are supported by a much wider evidence base (including, above all, written records) than any claims about prehistory. A version of the pessimistic induction that targets only claims about prehistory would not be self-defeating at all.

Thus, since there is no reason to think that the asymmetry of manipulability and the role asymmetry of background theories make any difference to the strength of the pessimistic meta-induction, we may conclude, if somewhat tentatively, that the best hypothesis is (a): the pessimistic induction has just as much force against historical realism as it has against experimental realism. Thus, the results of this chapter are largely negative. The analogue asymmetry does not confer any epistemic advantage on historical science, relative to experimental science, and the asymmetries of manipulability and background theories do not affect the strength of the pessimistic induction.

5

Novel predictions in historical science

Scientific realists sometimes respond to the pessimistic induction from the history of science by placing extra weight on predictive novelty. Sure, there are plenty of cases in which scientific theories yielded accurate predictions in their day, and were later shown to be mistaken. But were those predictions *novel*? Philosophers who favor the pessimistic induction will have a more difficult time producing examples of discredited theories that once enjoyed novel predictive success. What's more, contemporary scientific realists, such as Leplin (1997) and Psillos (1999) typically defend their views with a new and improved version of the traditional abductive argument for realism. According to this new version of the argument, the phenomenon that cries out for realist explanation is not just any old empirical success, but *novel predictive success*. Do the asymmetry of manipulability and the role asymmetry of background theories have any bearing on this new line of argument for scientific realism?

I believe that they do. The asymmetry of manipulability and the role asymmetry of background theories mean that novel predictive successes are harder to come by in historical than in experimental science.

5.1 NOVEL, UNTESTABLE PREDICTIONS

Suppose that some scientific theory T, when conjoined with a set of background assumptions A, predicts an empirical result O. And suppose that O is a novel prediction in the sense spelled out by Jarrett Leplin (1997) in his book-length defense of scientific realism. No other theory predicts O, so it satisfies Leplin's uniqueness condition, and the scientists who devised theory T did so without knowledge of whether O would obtain, so it also satisfies Leplin's independence condition. Suppose, in addition, that some other well-confirmed set of background assumptions *B* (which may or may not overlap with *T* and *A*) implies that scientists will probably never observe the result *O*. Interestingly, Leplin himself thinks that current theorizing in physics may provide us with examples that conform to these two suppositions. That is, current physics supplies examples of *novel, but untestable predictions*.

Leplin points out that "some of the latest physical theories are difficult, if not impossible, to test empirically" (1997, p. 178). One thing that can make a theory difficult to test is that it yields few or no predictions. But the problem here, according to Leplin, is different:

> It is not that current theories fail to yield empirical predictions. Nor is it that the predictions they yield fail to distinguish them from other theories already tested. Opportunities to meet conditions for novelty abound in current physics. The problem is that such predictions do not appear to be testable. We lack the technological means to determine whether they are correct. Nor is this problem reasonably regarded as a temporary limitation, comparable to the epistemic situation that has frequently confronted new theories, so that a "wait-and-see" attitude is appropriate. The new situation is that the very theories whose credibility is at issue themselves ordain their own noncomfirmability. If the latest theories are correct, then we should not be able to confirm them. (Leplin 1997, p. 178)

String theory affords an especially vivid illustration of Leplin's point (although see his 1997, chapter 7, for numerous other examples drawn from recent physics and cosmology). According to string theory, the tiniest subatomic particles arise as a consequence of the vibratory patterns of even tinier strings. The uninitiated person (like me) might wonder naively if scientists will ever be able to use particle accelerators to detect superstrings and measure their properties. In his popular book on string theory, Brian Greene writes that

> Without monumental technological breakthroughs, we will never be able to focus on the tiny length scales necessary to see a string directly. Physicists can probe down to a billionth of a billionth of a meter with accelerators that are roughly a few miles in size. Probing smaller distances requires higher energies and this means larger machines capable of focusing that energy on a single particle. As the Planck length is some 17 orders of magnitude smaller than what we can currently access, using today's technology we would need an accelerator the size of the *galaxy* to see individual strings. (Greene 1999, p. 215)

So apparently we will not be using particle accelerators to study the properties of strings anytime soon. Although it might be possible to test string
theory in other less direct ways, this passage from Greene's book gives
us a rough and ready idea of what it would mean to derive a *novel,
but untestable prediction*. Maybe physicists could predict what we would
observe if we had a particle accelerator the size of the galaxy, but we
obviously cannot test that prediction.

According to Leplin, the best argument for scientific realism is the
argument from novel predictive success, a member of the well-known
family of abductive arguments for scientific realism. The particular version of realism that he defends is minimal epistemic realism, or the thesis
that "there are possible empirical conditions that would warrant attributing some measure of truth to theories – not merely to their observable
consequences, but to theories themselves" (1997, p. 102). Realists argue
that that the novel predictive success of a theory would be inexplicable if that theory were not true or approximately true. Since he worries
that many of the novel predictions derived from the most fundamental
physical theories are untestable, Leplin is pessimistic about the prospects
for using the argument from novel predictive success to defend a realist interpretation of many of the claims that theoretical physicists make,
say, about strings. Indeed, he seems to concede that the pervasiveness of
novel, but untestable predictions in theoretical physics means that our
scientific realism should only extend so far:

> My own defense of realism is impotent in the face of this development. If
> this defense is the last word on the realist view of science – if, that is, novel
> success is the only basis for realist commitments (whether or not it is a basis
> at all) and my conditions for novelty are not only sufficient but necessary
> (which I have not claimed) – then the most fundamental theories of physics
> are not to be believed. Pragmatism, or less, becomes the right philosophy,
> for purely scientific reasons. (Leplin 1997, p. 182)

One can reasonably interpret Leplin here as saying that portions of current fundamental physics are at an epistemic disadvantage compared
to much of the rest of modern science. Novel predictive success is
the best sort of evidence that one can get in empirical science, but in
some cases the theories themselves, together with the fact that we cannot experimentally intervene in natural processes at the tiniest physical
scales, give us reason to think that such success is unlikely ever to be
achieved.

In this chapter, I take up a related question which Leplin does not himself discuss. To what extent, if at all, do historical sciences such as geology, archaeology, and paleobiology suffer from the same epistemic disadvantage? There may not be any straightforward answer to this question, but I hope to advance the discussion in the right direction by arguing for a few modest claims, in something of a dialectical progression:

(1) There are some fascinating examples of successful novel predictions derived from claims and theories about the distant past. This may be old news to some readers, but it is worth repeating loudly since one sometimes still encounters the view that historical scientists typically explain things by fitting them into narratives, and that one cannot derive predictions from narratives.

(2) On the other hand, though, the factors that make it so difficult to test novel predictions in the more speculative areas of current physics – and which lead Leplin to be so pessimistic about the future of scientific realism – are also very much in play in historical science.

(3) On the other hand, though, historical scientists in fields such as paleobiology have had some success in compensating for this epistemic disadvantage, mainly through the use of new technologies.

Thus, my aim in calling attention to the relative epistemic disadvantage that historical science shares with some work in current physics is not to denigrate historical science or to raise doubts about its scientific status. Instead, I wish to tell a more complicated story of historical science struggling to emerge from a position of relative epistemic disadvantage. This is a rather different view of historical science than the one recently defended by Cleland (2002), who sees historical and experimental science as being about equally well off, from an epistemic point of view, because both take advantage of different aspects of the asymmetry of overdetermination.

In the next section, I present one version of the realist argument from novel predictive success. Then in the remainder of the chapter, I go on to develop claims (1) through (3).

5.2 WHY SUPPOSE THAT PREDICTIVE NOVELTY CARRIES ANY EXTRA EVIDENTIAL WEIGHT?

In 1869, when Mendeleyev devised the periodic table of the elements, only sixty elements were known, and the table contained several gaps. According to the periodic law, several interesting properties of the elements are periodic functions of atomic weight. That is, when the elements are arranged in order of their atomic weights, certain properties (e.g. reactivity) recur at regular intervals. At first, no one in the scientific community thought that the periodic law was terribly important, since Mendeleyev had done little more than accommodate and organize the available data. But when other scientists independently discovered two of the elements whose existence he had predicted, the Royal Society bestowed upon him the prestigious Davy Award. As Peter Lipton puts it, "Sixty accommodations paled next to two predictions" (1991, p. 134). Notice, first, that Mendeleyev's predictions satisfy both of Leplin's conditions for predictive novelty. Nobody else had a theory that predicted the existence and properties of these two elements (so Mendeleyev's prediction was unique), and at the time he formulated the theory, neither Mendeleyev nor anyone else knew whether those two elements really existed (so his prediction was independent).

What if Mendeleyev had waited until after the discovery of gallium, scandium, and germanium to formulate his periodic law? Either way, the logical relations between the theory and the data are exactly the same. But realists hold that the evidence for the periodic law would have been weaker if Mendeleyev had waited until after the discovery of these three elements, and then used the information about them to formulate the periodic law.

The basic argument for this view is as follows: Suppose we remark that in both cases, the periodic law fits the empirical data extremely well. Indeed, the fit is equally good in both cases. But why does the law fit the data so well? What explains the goodness of fit? Suppose we reason abductively, as scientists often do: First, we generate a pool of potential explanations of the striking good fit between Mendeleyev's periodic law and the data. Then we determine which explanation is best according to some reasonable standard (simplicity, coherence, or something along those lines). Finally, we infer that the best explanation is probably true. Notice that if Mendeleyev waits until after the discovery of gallium, scandium, and germanium, we can explain the goodness of fit between the theory and the data by supposing that he tailored the theory to accommodate the data – that is, we can explain the goodness of fit by supposing that the theory is *ad hoc*.[1] Following Lipton (1991), we can call this the *tailoring explanation.* On the other hand, if Mendeleyev formulates the

---

[1] Psillos (1999, pp. 106–107) also stresses the connection between novelty and ad hocness, in the following way. He says that when some theory T predicts evidence E, the prediction can be "use novel" even if E was known at the time the theory was devised, so long as

periodic law before the discovery of those three elements, we cannot use the tailoring explanation, because even though Mendeleyev was a great scientist, he could not possibly have tailored his theory to fit nonexistent data. So how shall we explain the goodness of fit in that case? Realists argue that the goodness of fit would be a complete mystery if the theory of the periodic table were not true or approximately true, and if the entities and kinds described by the theory did not really exist. Novel predictive success entitles us to say that our theories deliver the truth (or at least, something close to the truth) about unobservables.

A number of critics have raised blanket objections to this line of argument. For example, if abduction, or inference to the best explanation, is an unreliable form of inference to begin with, then the inference to the best explanation of novel predictive success will not mark any significant improvement over its predecessors. For purposes of this chapter, I will assume that these blanket objections can be answered. This is the same strategy that I adopted in chapter 3: By assuming – though only for the sake of argument – that realists can handle the blanket objections, we can zero in on the question whether the strength and scope of the argument to the best explanation of novel predictive success are at all affected by the asymmetry of manipulability and/or the role asymmetry of background theories.[2]

One final bit of clarification is in order. Mendeleyev formulated his theory of the periodic table and predicted the existence and properties of gallium, scandium, and germanium. These three elements were then discovered at a later time. This case involves *temporal* novelty, in the sense that at the time the predictions were made, nobody knew whether the predicted facts would really obtain. The realists who have developed this line of argument like to point out, however, that while temporal novelty is sufficient for the realist argument to go through, it is not necessary. Indeed, realists typically deny that novelty is simply a matter of timing. In order to see why, we can perform some imaginative manipulations of the Mendeleyev case.

Suppose that unbeknownst to Mendeleyev, some other scientists, working independently, have already discovered the existence and properties of gallium, scandium, and germanium, though they know nothing about

the theory is not ad hoc. He then gives two conditions for ad hocness: Information about E must not have been used in the construction of T, and T must not have been revised solely for the purpose of accommodating E.

[2] For another interesting blanket objection to the argument from novel predictive success, see Horwich (1982, pp. 111–112).

the theory of the periodic table. When these other scientists first read about Mendeleyev's predictions they recognize instantly that those predictions are true. In this case, the predicted facts are known (by someone) even before Mendeleyev derives his predictions, but since Mendeleyev did not himself know those facts, we cannot invoke the tailoring explanation, and the realist argument goes through.

Next, imagine that Mendeleyev formulates his theory of the periodic table using only his knowledge of the sixty previously discovered elements. But while he is working on the theory, he learns that other scientists, working independently, have discovered the existence and properties of gallium, scandium, and germanium. So he knows about the existence and properties of these three elements before he derives his predictions. However, if we could show that Mendeleyev *did not use* his knowledge of the three new elements in forming his periodic law, then we still could not invoke the tailoring explanation, and the realist argument would still go through. If he did not use his knowledge of these facts in formulating the law, he could not have tailored the law to fit those facts. What this case shows is that predictive novelty is not at bottom a matter of timing at all; it is, rather, a more complicated epistemic notion having to do with what knowledge scientists do and do not use when formulating their theories. This last case exhibits what Psillos 1999 calls "use novelty." This is also what Leplin tries to capture with his independence condition.[3]

---

[3] To make this more precise, Leplin writes that the prediction of some observational result O from theory T satisfies the independence condition for novelty when "There is a minimally adequate reconstruction of the reasoning leading to T that does not cite any qualitative generalization of O" (1997, p. 77). A reconstruction is adequate when it gives us sufficient reason to propose the theory, to subject it to further tests, etc. At first, it seems like the point of such a reconstruction is to represent the actual reasoning of the theories: did the theorist, or did he not, in fact cite any qualitative generalization of O when reasoning his way to the theory? But which background knowledge a theorist in fact relied on at a given point is a psychological question, and Leplin repeatedly insists that his independence condition must be understood in a purely logical way:

> The idea is to circumscribe novelty by identifying what is essential to the reasoning used to generate the theory, specifically with respect to the role of empirical results. But again, "essential" is to be understood in a logical, rather than a psychological sense. It might be that the theorist would not, as a matter of psychological fact, have thought of the theory unless he knew of a certain empirical result, although the theory could have been validly deduced from other knowledge he had and assumptions he made. Then the result would not figure in a minimally adequate reconstruction (Leplin 1997, p. 72).

Leplin appears to want to say that even if Mendeleyev (for example) had in fact used his background knowledge of the three elements germanium, scandium, and gallium, in reasoning to the periodic law, the prediction of the existence and properties of those three elements would still be novel. Why? Because Mendeleyev had enough other background

Leplin (1997) also stresses the importance of what he calls the uniqueness condition. Suppose that another rival scientist has produced another theory which is quite different from and incompatible with the periodic law. The rival's theory, however, makes exactly the same predictions as Mendeleyev's, so that the discovery of the existence and the properties of gallium, scandium, and germanium does not discriminate between the two theories. So novelty involves uniqueness as well as independence.[4]

Once we realize that novelty is fundamentally an epistemic matter rather than simply a matter of timing, it is possible to dream up cases that will cause some problems for the scientific realist. One such thought experiment is due to James Ladyman:

> So, for example, if we found a dead scientist's revolutionary new theory of physics, but they left no record of what experiments they knew about or what reasoning they employed, it follows from the conditions above [i.e., Leplin's uniqueness and independence conditions] that such a theory could have no novel success and hence no amount of successful prediction of previously unsuspected phenomena would motivate a realist construal of the theory. (Ladyman 1999, p. 183)

Ladyman overstates his objection a bit, but it is worth examining because it brings out an important feature of the realist line of argument. Suppose that Mendeleyev knew of the existence and properties of gallium, germanium, and scandium when he derived his predictions. But suppose he died without leaving any record at all as to whether he used that knowledge of the existence and properties of the three newly discovered

knowledge at the time to mount an argument for taking the theory of the periodic table seriously and subjecting it to further testing, without having to mention those three elements. His knowledge of those three elements was "inessential" to the reasoning leading to the theory. I am inclined to think, in contrast with Leplin, that the notion of novelty that is needed for the realist argument to go through is fundamentally a psychological one. Fortunately this difference will not make any difference to the arguments about historical science that I develop in this chapter.

[4] Leplin writes that a prediction of some result O from theory T satisfies the uniqueness condition when

> There is some qualitative generalization of O that T explains and predicts, and of which, at the time that T first does so, no alternative theory provides a viable reason to expect instances. (Leplin 1997, p. 77)

A qualitative generalization of O is "the effect itself, the type of phenomenon – itself a type – that O instantiates, independently of considerations of quantitative accuracy" (Leplin 1997, p. 73). Thus if one geologist predicts that a glacier will recede by some measurement m in one year, and another, using a different model, predicts that the same glacier will recede further, by measure $m + n$, the predictions differ only quantitatively. Qualitatively, they are the same prediction, and so neither is unique, in Leplin's sense.

elements in formulating the theory of the periodic table. Ladyman's negative conclusion that in a case like this, the theory "could have no novel success" is not warranted. The problem is that we simply cannot tell whether the predictive successes in this case were novel, because we cannot tell whether the independence condition is satisfied. (This is just an example of local underdetermination in the history of science: hypotheses about what Mendeleyev did and did not use are underdetermined by the evidence available to later historians.) Nor should we rule out the possibility that someone else will come along at a later date and derive another prediction from the theory in question, and one that does clearly satisfy the conditions for novelty. What this shows is that in many cases – for example, in cases involving temporal novelty – it is easy to tell whether a given prediction is novel. But in other cases, knowing whether a prediction is novel will require a good deal of knowledge of the historical details.

5.3 NOVEL PREDICTIONS IN HISTORICAL SCIENCE

Some readers may well doubt that historical science ever yields any novel predictions at all. Some of the best recent discussions of the epistemology of historical science stress the notions of coherence (Thagard 2000; Kosso 2001; Tucker 2004), consilience, and/or explanatory unification (Gould 2002). According to the views of these writers, our main justification for believing what archaeologists, geologists, and paleobiologists tell us about the distant past derives not from empirical testing, but rather from what philosophers of science like to call "non-empirical" or "supra-empirical" theoretical virtue. Successful novel prediction, which is the very best sort of empirical virtue, does not play much of a role in these accounts of the epistemology of historical science.

In addition, some people find the following line of argument, which I will call the *argument from historical narrative*, persuasive:

P1. Historical scientists typically explain token events by fitting them into *narratives*, rather than by deriving them from statements of general laws plus initial conditions.
P2. But narratives, in contrast to general laws, do not yield any novel predictions.
C. Therefore, historical scientists typically do not make novel predictions.

Certainly P1 has found sympathy among writers such as Hull (1975); Gould (1989); Kitcher (1993); and Cleland (2002). The trouble with this line of reasoning, though, is that its soundness depends on how much, or how little, we choose to pack into the notion of "narrative."[5] Though I will not press the issue here, my own suspicion is that any conception of narrative which is restrictive enough to make P2 come out true will also make P1 false (and conversely, any conception which is expansive enough to make P1 seem plausible will make P2 come out false). Instead of exploring different analyses of narrative, my strategy will be to present some counterexamples that go a long way toward undermining the argument from historical narrative. I shall describe these counterexamples in some detail, because I want to dispel, once and for all, the myth that historical scientists never make novel predictions.

Paleobiology

Consider the reconstruction of the behavior and life habits of the cretaceous marine reptiles known as mosasaurs. Almost by accident, in the course of studying a mosasaur specimen that had a shark's tooth embedded in one of its vertebrae, Martin and Rothschild (1987, 1989) found that many mosasaurs suffered from the rare (in humans, at least) bone disease avascular necrosis. Avascular necrosis occurs when bone tissue dies as a result of a loss of blood supply. Martin and Rothschild found the telltale signs of bone liquefication due to avascular necrosis in a large number of mosasaur specimens collected from many different parts of the world, and from rock strata of different ages. Aside from physical injury, avascular necrosis in humans has only a few known causes, including radiation exposure and decompression syndrome, or the bends. The geographical and stratigraphical distribution of affected mosasaur specimens, together

---

[5] Interestingly, the claims that historians typically explain things in terms of narratives, and that one cannot derive predictions from narratives also crop up in a debate about the nature of historical explanation that took place during the 1950s and 1960s, but which subsequently petered out. On the one side of that debate, Carl Hempel argued that historical explanations are really only disguised or incomplete D-N explanations, which he called "explanation sketches" (1965, p. 238). On the other side of this debate, W. B. Gallie (1959) argued that history has its own distinctive kind of explanation, which he called "genetic explanation." According to Gallie, historians explain things by citing "temporally prior necessary conditions," though they are almost never in a position to give a more complete explanation in terms of sufficient conditions. But an explanation in terms of prior conditions that are (at most) causally necessary for the occurrence of some later event cannot support any predictions. Thus, Gallie (1959) may be an important source for P2 in what I am calling the argument from historical narrative.

with the fact that other creatures living at the same time did not suffer from avascular necrosis, seem to rule out radiation exposure as a cause. Martin and Rothschild argue that the mosasaurs must therefore have suffered from the bends: on occasion they dove too deep – whether in pursuit of prey or in flight from sharks – and ascended too rapidly. While avascular necrosis in humans can cause bones to lose their structural integrity, and can therefore be a painful and debilitating disease, its effects on marine creatures are less severe because their bodies do not need to withstand the same compressional forces as terrestrial animals must endure.

Martin and Rothschild also observed that the incidence of avascular necrosis varies across mosasaur genera: The smaller *Clidastes* showed no sign of the bone pathology. Every specimen of *Tylosaurus* that the scientists examined had suffered from avascular necrosis. The genus *Platecarpus*, on the other hand, had on average the largest number of affected vertebrae per individual. In order to make sense of the data, they floated the hypothesis that the earliest mosasaurs were surface feeders, but that later mosasaurs like *Tylosaurus* and *Platecarpus* took to deep-sea diving.

Sheldon (1997) set out to test this hypothesis about diving behavior by looking at cross-sections of mosasaur rib bones in order to determine their density and structure. Sheldon notes that bone density is related to diving behavior because it affects an animal's neutral buoyancy. Animals with dense bones tend to have a shallow neutral buoyancy, whereas many living marine mammals that are capable of deep dives have less dense bones. Citing earlier work on the biomechanics of swimming, Sheldon argues that animals with less dense bones are capable of more energy-efficient dives. With this background knowledge, it is possible to derive a prediction from the Martin-Rothschild hypothesis about mosasaur diving behavior: *Clidastes*, the alleged "surface sculler," should have somewhat denser bones than the deep diving *Tylosaurus* and *Platecarpus*. This prediction satisfies the independence condition, because Martin and Rothschild formulated their hypothesis without knowing anything about variations in mosasaur bone microstructure. It also satisfies the uniqueness condition, because no other available theory of mosasaur ecology and behavior makes the same prediction.

When she looked a bone cross-sections under a microscope, Sheldon found just the opposite of what the Martin-Rothschild hypothesis would lead one to expect: The rib bones of *Platecarpus* are significantly more dense than those of *Clidastes* (9.4% porosity *vs.* 41.8%). *Platecarpus* even appears to have had another bone pathology known as *pachyostosis*, "which is an increase in bone density across the entire bone" (Sheldon 1997, p. 342). On the other hand, *Tylosaurus* had an even lower bone density than *Clidastes* (62.3%) – so low, in fact, that the animals probably had osteoporosis. The prediction derived from Martin's and Rothschild's hypothesis about mosasaur behavior came out wrong, and Sheldon was compelled to offer a more complex interpretation of these observations concerning bone density and bone pathology.

Archaeology

Although Peter Kosso (2001) does not explicitly discuss the subject of novel prediction in his recent book on the epistemology of archaeology, he does describe an interesting example of novel prediction in that field. During the 1980s, the Southern Euboea Exploration Project (SEEP) set out to test the hypothesis that Athens had maintained a cleruchy on the Paximadhi peninsula, near the town of Karystos in southern Euboea, beginning around 450 and continuing until around 400 BCE, when the Athenian empire fell apart. A cleruchy was a special kind of settlement whose members (usually Athenians from the poorer classes) were given land to farm in exchange for military service. A cleruchy, in short, was something of a hybrid between a garrison and a colony. The date of 450/449 BCE was arrived at by looking at the records of the yearly tribute which the town of Karystos paid to Athens. In 449, the amount of tribute declined (while the tribute paid by other towns remained constant), suggesting that the value of the property under the control of the town had declined – just what one would expect if a portion of the property had been seized for the establishment of a cleruchy.[6] Archaeologists also knew from an earlier survey that there were farms on the Paximadhi Peninsula that could be dated roughly to the classical period. So in order to test the cleruchy hypothesis, they derived a number of predictions about what a more extensive study of the area would reveal. (1) Since written records from Diodorus Siculus and Plutarch suggest that the Athenians sent about 250 cleruchs to Eoboea, a careful survey should reveal approximately that number of farmsteads. (2) The layout and organization of the

---

[6] Since some of the background evidence used in this case involves historical records, this should be classified as an instance of historical rather than prehistorical archaeology. Nevertheless, it is still a vivid example of the derivation of a novel and testable prediction from a hypothesis about the unobservable distant past. And crucially, the predicted results do not involve written testimony.

network of farms on the Paximadhi Peninsula should resemble that of farmsteads that have been found in Attica. (3) Pottery remains at the site should consist mainly of coarseware – the sort of pottery that lower-class Athenians would have used for daily chores. (4) Precise dating techniques should show that the site was inhabited between 449 and about 400 BCE. These predictions all exhibit novelty, in Leplin's sense. For example, by dating the garbage found at the very bottom of a cistern, archaeologists can determine when people stopped using the cistern to store water. The prediction would be that the inhabitants of the Paximadhi site stopped using their cisterns for water storage around 400 BCE. No other available hypothesis makes just this set of predictions, and the archaeologists associated with the SEEP considered these predictions before they undertook a more careful survey of the site. In this case, the predictions turned out to be mostly accurate.

Historical geology

One of the more controversial theories in contemporary geology is the "snowball Earth" theory, according to which the entire planet froze over on several different occasions toward the end of the Proterozoic era, between 800 and 500 million years ago, with each snowball Earth episode lasting several millions of years. Proponents of the snowball Earth theory point out that it can explain a variety of otherwise puzzling geological phenomena, including (1) glacial debris in locations that would have been in the tropics back in the neoproterozoic when the rocks were formed; (2) thick layers of carbonate rock that sit right on top of the glacial debris; (3) unusual ratios of carbon isotopes in neoproterozoic rocks, which suggest that photosynthetic activity in the earth's oceans came to a halt, and (4) iron-rich deposits in neoproterozoic rock (Hoffman and Schrag 1998, 2000). Though no one denies the explanatory power of the snowball Earth theory, the theory is vulnerable to the criticism that it has been tailored to fit much of this empirical evidence. In the last few years, however, some geologists have tried to test the snowball Earth theory by deriving novel predictions from it and testing those predictions out in the field. For example, Leather, et al. (2002) studied a portion of the Ghadir Manqil Formation, in northern Oman. They knew on the basis of previous work that this is one of several places on earth where one can find neoproterozoic glacial debris. They set out to test the snowball Earth theory vis-à-vis the hypothesis that this particular glacial debris had been deposited by a series of advancing and retreating glaciers similar to those that covered much of the northern hemisphere during the Pleistocene. If the snowball Earth hypothesis were true, one would expect to find evidence of "hydro-logic shutdown," or long periods of several million years during which no sedimentary deposits were formed. Instead, Leather, et al. found that at the relevant sites in Northern Oman, bands of glacial debris were inter-spersed with and broken up by layers of sediment that were probably deposited by floods, and that several of these glaciation/flooding cycles occurred over a relatively short period of time. We can interpret Leather, et. al. as testing a novel prediction about what the rocks of the Ghadir Manqil formation would look like, if the snowball Earth theory were true. No theory other than the snowball Earth theory predicted that the sites in Oman should reveal evidence of hydrologic shutdown, and the researchers in this case surveyed the sites in Oman in order to test this prediction.

These examples show unequivocally that scientists do sometimes derive novel predictions from claims about the distant, unobservable past. This is good news for those who wish to defend realism about the past. For they can argue that when such predictions come out right (as in the case of the Athenian cleruchy), it would be surprising indeed if the the-ory or hypothesis from which the prediction is derived were not true or approximately true. In the cases above, the novel predictions were not only testable, but actually were tested.

5.4 WHY ARE NOVEL PREDICTIONS IN HISTORICAL SCIENCE SO DIFFICULT TO TEST?

So far, I have only argued that the view that historical scientists never make any novel predictions at all is overly pessimistic. The counterexam-ples to this overly pessimistic view seem to paint a rosy picture by contrast. Not only do historical researchers derive novel predictions from claims about the past, but in the cases described, they also succeeded in testing those predictions. If those examples are typical of historical science, then historical science would seem not to be at any epistemic disadvantage vis-à-vis experimental science.

But not so fast. We should be cautious about accepting this rosy con-clusion, because there are three factors which make it relatively more difficult to test novel predictions in the context of historical science: (1) First, historical processes – such as the movement of tectonic plates, or macroevolutionary processes – occur very slowly, relative to our life spans. This means that although we might be able to make predictions based on observed trends about where those processes will go in the future, we will not be around long enough to check to see if those predictions are born out. (2) A great many predictions in experimental science have the form: "If we do thus-and-such, under these conditions, and if this theory is correct, then that is the effect we should observe." That is, scientists predict what will happen as a result of certain experimental manipulations. But since we cannot manipulate the past (the asymmetry of manipulability again), our powers of testing predictions derived from claims about the past are limited. (3) In some cases, historical scientists possess background information that gives them independent reason for doubting that a result predicted by some historical theory will ever be observed. This can happen when scientists are working with background theories that tell us how historical processes destroy crucial bits of evidence (the role asymmetry of background theories again). These three considerations do not mean that historical scientists never make any novel predictions; nor do they imply that novel predictions derived from theories and hypotheses about the past are never empirically testable. For we have just seen some counterexamples to both of these sweeping claims. But these considerations do have important consequences. They give us good reason to think that novel predictive successes in historical science will be fewer and further between than similar successes in experimental science. If this is correct, it means that historical science is, once again, at somewhat of an epistemic disadvantage with respect to experimental science. In the remainder of this section, I provide illustrations of the three considerations above by presenting a series of cases in which novel predictions turned out to be untestable. If I am right, then historical science is in somewhat the same predicament as fundamental physics when it comes to checking to see if novel predictions are true.

(1) *Problems of scale.* The theory of plate tectonics yields a large number of predictions that are novel by Leplin's standards but untestable (at least by us), and hence not at all risky. The reason for this is simply that the processes involved take too long. At Pinnacles National Monument, in California, the rocky spires that inspired someone to give the park its name consist of volcanic rock that is not found anywhere else in California's central coastal range. Geologists think that the "pinnacles" are actually the remains of a volcano that erupted about 23 million years ago, some 195 miles to the southeast of the present location of the park,

in what is now the Los Angeles basin.[7] The resulting volcanic mountain was carried gradually northward as the Pacific plate ground along the San Andreas fault. In the meantime, processes of weathering and erosion removed much of the original mountain, leaving only the exposed rocks that attract climbers to the park today. This story about how Pinnacles National Monument came to have its distinctive features is a classic example of a historical narrative built around what David Hull (1975) would call a "central subject," or an individual which persists and undergoes various changes through time. In this case, the central subject is the mountain itself. One might think that such a story cannot serve to generate any predictions, and the argument from historical narrative, considered earlier, would seem to lend some support to this intuition. However, while there is an interesting methodological problem here, someone who simply denies the possibility of deriving novel predictions from narratives has misidentified that problem. In fact we can derive a prediction from this theory about the origins of the pinnacles: Assuming that the tectonic plates continue moving at about the same rate, in another 23 million years, whatever is left of the original volcanic mountain will have moved another 195 miles to the northwest (a location that is currently off the coast of northern California, not far from San Francisco). The problem is that the processes involved take so long that we will not be around to test such predictions, and this is one way in which predictions can fail to be risky.

(2) *Problems of manipulability.* In addition to reconstructing extinct organisms, paleobiologists also seek to test claims about large-scale trends in evolution, and they do this by looking at the "big picture" presented by the fossil record – that is, at the patterns of speciation and extinction in different clades. Over the last few decades, paleobiologists have developed ever more sophisticated methods for testing models against the data from the fossil record. Even so, this model testing is hampered by their inability to subject the processes in question to experimental manipulation. Here the impossibility of experimental manipulation has partly to do with the fact that the processes occurred in the past, and partly with the fact that the macroevolutionary processes are just too long and drawnout (as in the foregoing example). Consider, by way of illustration, the recent and ongoing debate concerning models of increasing biodiversity over time. One of the central claims of Darwinian evolutionary theory is

---

[7] Information about the history and prehistory of Pinnacles National Monument is available online at www.nps.gov/pinn/pphtml/nature.html, last accessed on August 30, 2005.

that if we go far enough back into the past, we will find that all the many millions of species that exist on earth today have descended from some common ancestor. However, as Benton (1997) points out in his helpful introduction to the debate, there are at least two different ways of getting from one species to the millions that we find today. First, it could be a general rule that species tend to increase in number exponentially over time, without ever reaching any equilibrium. According to this expansion model, the history of life on earth has been one of exponential growth in the number of species, interrupted only by a number of mass extinction events. A second possibility, the logistic model, is that diversification takes the form of an S-curve, with an initial period of slow increase followed by an explosive increase in the number of species, after which the rate of increase slows again as the number of species reaches an equilibrium where the extinction rate and the speciation rate balance each other out. Notice that the overall rate of increase or decrease in the number of species will vary depending on the speciation and extinction rates. According to the expansion model favored by Benton (1995), the number of species increases at an exponential rate, and this rate of increase does not vary depending on how many species there are. According to the logistic model favored by Sepkoski (1978, 1979, 1984), by contrast, the rate of increase of the number of species does depend on how many species there are: as the number of species approaches the ecological carrying capacity, the extinction rate increases relative to the speciation rate, and the rate of increase in the overall number of species slows until it approaches equilibrium.[8]

For present purposes, the interesting questions are, first, whether it is possible to derive any novel predictions from these two models, and second, whether those predictions are empirically testable. In addressing these questions, I want to begin by making a fairly obvious but important point: From each of these models, it is possible to derive novel predictions about what would happen if we were to perform certain experimental manipulations. The trouble is that we cannot and will never be able to perform those manipulations, because we cannot intervene in the past. Stephen Jay Gould famously described the consequences of our inability to replay the tape of evolution:

---

[8] Cuddington and Ruse (2004) argue that Sepkoski's commitment to the logistic model, even though the empirical issues are not quite settled, reflects a deeper *a priori* commitment to the idea of the balance of nature. More generally, one might think that difficulties of empirical testing leave more room for scientists' views to be influenced by non-empirical considerations.

> We live, as our humorists proclaim, in a world of good news and bad news. The good news is that we can specify an experiment to decide between the conventional and the radical interpretations of extinction, thereby settling one of the most important questions we can ask about the history of life. The bad news is that we can't possibly perform the experiment.[9] (Gould 1989, p. 48)

Try to imagine a long-lived super-experimenter who is actually capable of rewinding the tape of evolution and playing it back – perhaps varying some of the initial conditions while holding others constant. (Or perhaps the super-experimenter can look at two systems, Earth and Twin Earth, that are exactly alike but for one variable.) We can use the expansion model and the logistic model to make conflicting novel predictions about how the history of life would unfold, and the super-experimenter could just sit back and watch to see which predictions are born out. These predictions about what the super-experimenter would observe are, I submit, novel but untestable. Suppose, for example, that the super-experimenter decided to play back the tape of evolution while eliminating mass extinction events in order to see whether speciation and extinction rates really do arrive at an equilibrium, as predicted by the logistic model. These predictions, derived from the respective models, concerning what would happen on a macroevolutionary scale if certain conditions were varied, would be both independent and unique. But obviously *we* cannot test them. This is untestability due to unmanipulability. In cases like this one, where the untestability of the predictions is so obvious, one will not find scientists discussing the predictions as predictions at all; instead, they will sensibly tend to ignore them. There may well be *lots* of novel predictions derivable from historical theories, models, and hypotheses that fall in this category: they are so obviously untestable, owing to our inability to manipulate the past, that they are scarcely worth mentioning.

The fact that scientists cannot test novel predictions derived from these models when the predictions concern experimental interventions that they could not possibly conduct does not mean that there is no other way to subject the models to empirical test. But it does mean that subjecting them to empirical test has been complicated and messy, without any clear instances of novel predictive successes or failures. As it happens, the only way to test the models is to see which one best fits the available

---

[9] In the context of this discussion, Gould, like Sepkoski, is interested in challenging the expansion model. But Gould is not exactly defending the logistic model, either. The "radical interpretation" alluded to in this passage involves an understanding of the relationship between biological diversity and disparity.

data from the fossil record. One problem that has arisen is that it is possible to tinker with the models until they fit the data reasonably well. For example, Sepkoski (1978, 1979) argued that a logistic model best fits the record of diversification of marine fauna during the phanerozoic.[10] But the data from the fossil record of marine bivalves do not fit Sepkoski's early logistic model perfectly. To eliminate some of the discrepancies, Sepkoski (1984) later introduced a more complicated "tri-phasic" logistic model, which has three different groups of marine creatures diversifying according to three different curves. He divided marine animals of the phanerozoic (the last 550 million years or so) into three different fauna: Cambrian fauna (trilobites and others), Paleozoic fauna (cephalopods, ostracods, and others) and Modern fauna (bivalves, gastropods, malacostracans, fish, and others). Although the pattern of diversification of no single group takes the form of a clear S-curve, when Sepkoski looked at the total pattern resulting from superimposing these groups one over the other, he found that the resulting pattern fit the prediction of the logistic model extremely well. But the crucial point, for present purposes, is that he introduced the three-phase logistic model only after noting that other logistic models did not fit the available data quite as well. This type of adjustment of the model to the data, while not necessarily a bad thing, is a far cry from the risky testing of novel predictions, and one could easily invoke the tailoring explanation to make sense of the goodness of fit between the model and the data achieved by Sepkoski (1984).

There are other complications as well. For present purposes, though, I merely want to stress that the empirical work with these models has had more to do with accommodating the data in the fossil record than with testing novel predictions. Although it is possible to derive novel predictions from these models, some of those novel predictions would only be testable if we could manipulate the past and run macroevolutionary experiments.

(3) *Problems arising from the destruction of historical traces.* Finally, there is a range of cases in which background theories give scientists some reason to think that a predicted result will never actually be observed. Consider Darwin's reasoning that if any two living species are descended, with modifications, from some common ancestor, then there ought to be remains of intermediate forms in the fossil record. This prediction was

---

[10] For a highly accessible introduction to Sepkoski's work in this area, as well as a discussion of the values that inform that work, see Ruse (1999, chapter 11).

unique, in the sense that no non-evolutionary theories of the day predicted anything similar. It was also independent, since at the time, no clear-cut examples of transitional forms had been identified. In the end, this also turned out to be a success story; the prediction was confirmed with the discovery of *Archaeopteryx*, an apparent transitional form between reptiles and birds, in the 1860s. Early on, though, Darwin's critics seized upon the fact – of which Darwin himself was well aware – that this prediction, however novel, was not terribly risky. Indeed, if no intermediate forms turned up, Darwin and his followers could always chalk that up to the incompleteness of the fossil record, and for awhile that is exactly what they did. It was not clear in this case just what would have counted as a predictive failure. One immediate lesson to be drawn from this is that novelty alone does not necessarily make a prediction risky. Of course, we could have drawn that very same lesson from the earlier example of the galaxy-sized particle accelerator: Physicists might be able to predict what we would observe if we had a particle accelerator that big, and the prediction would be novel, but it would not be accompanied by the slightest degree of risk.

For a more vivid illustration of this point, consider once again the example of the titanosaur tracks from chapter 1. Wilson and Carrano (1999) wanted to test the hypothesis that the titanosaurs made wide-gauge sauropod trackways. As they themselves point out, we can easily derive the following prediction from that hypothesis: The discovery of a titanosaur skeleton at the end of a wide-gauge trackway would settle the issue. This prediction satisfies the independence condition, because no one has yet found such a specimen. It also satisfies the uniqueness condition, because the only rival hypothesis – i.e, that the brachiosaurs or the diplodocids made the wide-gauge tracks – yields a very different prediction. However, our background theories of taphonomy tell us that the conditions which are conducive to the preservation of skeletal remains are not at all conducive to the preservation of footprints. Background theories give us very good reason to believe that no one will ever find a titanosaur skeleton at the end of a wide-gauge trackway. The impossibility of subjecting this novel prediction to any serious empirical test is part of what motivated Wilson and Carrano to look instead at titanosaur hindlimb morphology, as described in chapter 1.

Incompleteness of the fossil record also creates problems for paleo-biologists' attempts to test predictions derived from the logistic and the expansion models of biological diversification. The problem is that the fossil record is more complete at the level of families than at the species or genus level. Much of the relevant work in this area has focused on families, because that is the highest resolution paleobiologists can achieve (or as far downward in the taxonomic hierarchy they can go) before the incompleteness of the fossil record becomes a serious problem.[11] However, there is no *a priori* reason why increases and decreases in the number of species must track increases and decreases in the number of families. The number of families could remain constant over time while the number of species waxes and wanes. The reverse could happen, too: Imagine a scenario in which the number of species in one particular family increases rapidly, while a number of other families, each containing only a few species, go extinct. Suppose that the loss of species in those other families is perfectly balanced by the gain of species in the first family. Then we would have a situation in which the number of families decreases while the number of species remains the same. What this means is that even if the expansion model (say) fails to fit the data very well at the family level, proponents of that model may still argue that the rate of increase in the number of species is likely to have been exponential (Benton 1997). In short, the incompleteness of the fossil record makes the testing of predictions derived from the expansion and/or the logistic model a much less risky prospect than it would otherwise be.

Finally, it will be instructive to return to the example of the snowball Earth hypothesis. Geologists have long puzzled over clear evidence of low-latitude glaciation during the neo-proterozoic, about 800 to 580 million years ago. But how could there be glaciers at low elevations near the equator? Presently the only glaciers near the equator occur at very high elevations (for example, on the slopes of Mt. Kilimanjaro). Back in chapter 2, I mentioned that two distinct hypotheses have been introduced to explain the neoproterozoic glacial debris: the snowball Earth hypothesis and the high obliquity hypothesis. According to the former, the entire earth was covered by an ice pack on several different occasions during the neoproterozoic, for several million years at a time. According to the latter, the angle of the earth's axis was different during that time period, so that the equatorial regions were much colder while the polar regions received more sunshine than they do today. In chapter 2, I used this case as an example of a local underdetermination problem in historical science, but

---

[11] For an elegant presentation of the arguments for focusing on family level data as opposed to species level data, see Huss (2004, pp. 178ff.).

it could just as easily serve as an example of an untestable, novel prediction. (Either way, the role asymmetry of background theories is the source of the trouble.) The snowball Earth hypothesis, but not the high-obliquity hypothesis, predicts that there should be evidence of glaciation in neoproterozoic rocks from high latitudes as well as from low latitudes. This prediction is unique, because no theory other than the snowball Earth scenario predicts the occurrence of glacial debris in high-latitude as well as low-latitude rocks. It is also independent, because at the time the snowball Earth hypothesis took its present shape, during the early 1990s, no one had done a systematic survey. But Evans (2000) noted that although it is not too difficult to identify glacial debris as having been formed during the neoproterozoic – and lots of examples have been studied, on several different continents – it is much more difficult to determine where the glacial debris was formed during the neoproterozoic. The only way to get a fix on where the glacial debris was deposited (i.e. whether in high latitudes or in low latitudes) is to use paleomagnetic evidence. Evans (2000) looked at all the neoproterozoic glacial deposits for whose location we have reliable paleomagnetic evidence – there are not very many – and found no evidence of high-latitude glacial deposits. This is another case in which a failure to observe a predicted result can be attributed to the incompleteness of the geological record. The basic problem that Darwin encountered when he predicted the occurrence of transitional forms in the fossil record is a recurring theme.

Now to pull everything together: My purpose in going into so much detail with the examples in this section is to drive home the point that although historical researchers can and do sometimes derive testable novel predictions from theories and hypotheses about the distant past, the three factors discussed here – problems of scale, our inability to manipulate the past, and the fact that we know that historical processes often destroy the evidence – raise special problems for the testing of novel predictions in the context of historical science. Taken together, these three factors mean that untestable, novel predictions will be quite common in historical science, as they are in fundamental physics. And since one cannot have novel predictive success without testable novel predictions, this also means that novel predictive successes will be somewhat fewer and further between in historical than in experimental science. Just how few and far between they will be is not something that I will try to settle here. The important thing is that this, in turn, means that historical science is (once again) at somewhat of an epistemic disadvantage with respect to experimental science.

5.5 COPING WITH THE ASYMMETRIES

Researchers in the fields of paleobiology and geology are aware of these problems of scale, incompleteness, and manipulability, and they have developed some interesting strategies for coping with them. Indeed, the best way to understand the recent history of paleobiology and geology is as a series of attempts to overcome these problems of empirical testing.

I will now present two case studies, one from paleobiology and one from geology, in which scientists have invented new tools to compensate for the asymmetry of manipulability and the role asymmetry of background theories. My thesis is that these recent developments need to be understood as attempts by historical researchers to cope with a situation of relative epistemic disadvantage. The case studies involve two revolutions that have occurred over the last few decades: The revolution in molecular genetics and the revolution in computing technology. And both involve the checking of novel predictions that would have been utterly untestable but for the new technology.

To begin with, one central problem of historical biology is that of phylogenetic reconstruction. The fact that existing species, as well as species represented only by fossilized remains, are related to one another by common descent makes it reasonable to ask, with respect to any three species, which is more closely related to which. This sort of question frequently arises in the context of discussions of human evolution. Take, for example, (A) modern humans, or *H. sapiens*; (B) modern chimpanzees, or *Pan troglodytes*; and (C) a species known only from fossil remains, such as *Australopithecus afarensis*. It is reasonable to ask whether (A) or (B) is more closely related to (C). Using the language of cladistics, we can define a monophyletic group (or a clade) as a set containing a species and all of the species descended from it. Then we can ask whether (A) and (C) form a monophyletic group to which (B) does not belong, whether (B) and (C) form a monophyletic group to which (A) does not belong, and so on. Over the last several decades, cladists have developed sophisticated methods for tackling such questions.[12] At first, those methods involved looking at the phenotypic features of the organisms in question, but more and more, scientists have begun to bring DNA evidence to bear on problems of phylogenetic reconstruction. To give just one example, currently the most endangered species of canid on the planet is the Ethiopian wolf (*Canis*

---

[12] For an extended discussion of those methods, see Sober (1988).

simensis). Only a few hundred individuals remain today in the highlands of Ethiopia's Bale National Park, and the wolves are threatened by habitat loss and hybridization with local domesticated dogs. Studies of the mitochondrial DNA of the Ethiopian wolves have revealed that the species is actually more closely related to the gray wolves and coyotes of Europe and North America than to the local (and much smaller) African Jackals (Gottelli, et al. 1994).

Until very recently, it seemed that the use of DNA evidence for purposes of phylogenetic reconstruction would be limited to cross-species comparisons of living organisms. However, Noonan, Hofreiter, Smith, and colleagues (2005), changed all that by extracting genomic DNA samples from fossilized remains – one from a tooth, and one from a bone – of giant cave bears discovered in two caves in Austria. Using radiocarbon dating, the scientists determined that the remains were slightly more than 40,000 years old, whereas previously the oldest samples of genomic DNA obtained from fossilized remains were from well-preserved specimens under 20,000 years old recovered from permafrost or desert environments. One of the big problems confronting any attempt to obtain usable DNA fragments from fossils is that what was left of the cave bear DNA was mixed together with all sorts of microbial DNA as well as DNA from other living things that had occupied the cave environment in the meantime. In order to handle this problem, the scientists created two metagenomic libraries consisting of copies of all the genetic material obtained from the tooth and bone, respectively. They then used a computer program to check all the material in these libraries against the genomes of living dogs, since dogs and bears are close phylogenetic relatives. Where they obtained hits, they assumed that they had isolated bits of genetic material from the cave bears. They then checked the cave bear DNA against the genomes of living black bears, polar bears, and brown bears. An earlier study using mitochondrial DNA had suggested that living polar bears and brown bears are actually more closely related to the extinct cave bears than they are to living black bears. We can interpret Noonan, Hofreiter, and Smith, et al. (2005) as testing a novel prediction derived from this earlier phylogenetic reconstruction, where the prediction was simply that the genomic DNA evidence would yield the same phylogenetic results as the morphological and mitochondrial DNA evidence. In this case, the scientists used new technology (the creation of metagenomic libraries) to test a novel prediction derived from a phylogenetic reconstruction – that is, from a hypothesis about past evolutionary relationships – that would otherwise have been untestable. I say "one can interpret" because the scientists themselves have interpreted their work somewhat differently. They take the positive results more as a vindication of their new methods than as a confirmation of the earlier reconstruction.

5.6 NUMERICAL EXPERIMENTS

If the problem is that historical processes destroy data that we might wish to have in order to check to see if novel predictions derived from historical theories and hypotheses are true, then one potential solution might be to use technology (in the above case, gene sequencing technology) to extract new kinds of evidence from the fossil and geological records. On the other hand, if the problem is that we are prevented from testing novel predictions by the fact that we cannot manipulate past events or processes as we would like, then one possible solution is to use technology (in this case, computing technology) to create numerical models, or simulations of those processes which we *can* manipulate. With one notable exception (Huss 2004), this technique of numerical experimentation has not received much attention from philosophers of science. In this section, I argue that numerical experimentation is best understood as a technique for coping with the asymmetry of manipulability: If you cannot have real manipulability, then virtual manipulability is the next best thing.[13]

John Huss offers the following helpful characterization of the role of numerical experiments in historical science more generally, and in paleobiology more specifically:

> In essence, numerical experiment is a type of non-empirical testing, and has a self-contained quality that differentiates it from empirical tests. In numerical experiment, the computer model is not itself being tested for its agreement with nature and is not a candidate for acceptance or rejection. Instead, the ability of the behavior of the model to represent the behavior of nature in the relevant respects is assumed. Numerical experiment thus relies on internal comparisons: between model inputs and outputs,

---

[13] Since numerical experiments do not actually involve the manipulation of past processes, one might reasonably wonder whether it is legitimate to call them "experiments" at all. Huss (2004) does so on the grounds that they "have design features that have been devised to manage error, artifact, and conscious and unconscious bias in a way that allows for reliable inference from effects to causes" (p. 161). That is, he argues at length that numerical experiment involves experimental reasoning, in which one runs numerous trials with the aim of eliminating false positives and false negatives.

between model outputs generated under different conditions, between model outputs at various stages of the simulation process, between model outputs generated using versions of the model that differ in some specified way, et cetera. These results speak directly only to the behavior of the model(s) used in simulation, and to the wider empirical world only indirectly . . . (Huss 2004, p. 153)

Huss proceeds to describe in great detail some of the ways in which numerical experimentation has figured in recent paleobiological debates. For example, during the 1980s, David Raup and Jack Sepkoski analyzed data on extinctions from the *Compendium of Fossil Marine Families* (Sepkoski 1982) and found that extinctions peak about every 26 million years (Raup and Sepkoski 1984, 1986).[14] However, the data that Sepkoski and Raup looked at were all at the family level, and this left them open to the objection that the 26-million-year periodicity was just an "artifact" that would go away if only we could look at patterns of extinction at the species level. One more specific problem with Raup and Sepkoski's work was that many of the families they looked at had been identified using traditional methods of taxonomy. Two cladists, Patterson and Smith (1987, 1989) objected that most of the families used by Sepkoski and Raup were paraphyletic groups, and that if we restricted our attention to clades – that is, to monophyletic groups, or to groups consisting of a species and all of its descendants – we would not see the 26-million-year periodicity. Although the debate involved other issues as well, one of the main points of contention was whether families identified using the methods of traditional taxonomy, or families identified using cladistic methods would best represent patterns of extinction at the species level. Sepkoski and Kendrick (1993) used numerical experiments to try to settle this issue. Which method of classifying families – traditional or cladistic – will give us the best reflection of what is going on at the species level?

Sepkoski and Kendrick (1993) used a computer simulation that employs an algorithm to generate an evolutionary tree.[15] The algorithm is straightforward: Start with one lineage. Then with each unit of time, that

---

[14] See Benton (1999) for a helpful introduction to some of the problems with using such databases.

[15] As Huss points out, this model was a more sophisticated version of what is known as the "MBL model," a stochastic model of evolution that was developed during the early 1970s by a group of paleobiologists who met at the Marine Biological Laboratory. This group included David Raup, Stephen Jay Gould, Thomas J. M. Schopf, and Daniel Simberloff, as well as Sepkoski. See Huss (2004, chapter 1) for a detailed discussion of the MBL model and its development.

lineage has a certain probability of (a) going extinct; (b) persisting into the next time unit without branching, or (c) persisting into the next time unit while branching to form a new lineage. This algorithm will produce different evolutionary trees with each run. Next, they used two different algorithms to group the lineages formed by one run of the simulation into taxa (or families). One algorithm grouped the lineages according to cladistic methods; the other was designed to represent the methods of traditional taxonomy. This makes possible the following test: on a single run of the simulation, one can look at two things: first, how well the number of clades correlates with the number of lineages with each time unit, and second, how well the number of traditional taxa correlates with the number of lineages at each time unit. One can then check to see which correlation is stronger. Another clever trick that Sepkoski and Kendrick employed was to simulate the incompleteness of the fossil record by giving each lineage a certain probability of being sampled during each unit of time. Each sampling event was supposed to represent the discovery of a fossil by scientists. The traditional and cladistic classifications were then based not on the total information about the evolutionary tree produced by the computer program, but rather on this incomplete sampling of the underlying lineages. When they thus modeled the incompleteness of the fossil record, Sepkoski and Kendrick found that counts of traditional taxa correlated just about as well with counts of actual lineages as did the counts of clades. This went some way toward rebutting the criticisms of Patterson and Smith.

Aside from the paleobiological models that Huss discusses, modeling in paleoclimatology affords another good example of numerical experimentation.[16] The debate within the geological community concerning the snowball Earth hypothesis has been fueled at times by increasing sophistication in the use of climate models. That debate began back in the 1960s when geologists first noticed glacial deposits from the neoproterozoic. Paleomagnetic data seemed to indicate that the glacial deposits had been formed near the earth's equator. It does not take too much imagination to realize that one way of explaining glacial deposits in the tropics is to suppose that the entire planet was once covered in ice. Perhaps the first major obstacle facing the snowball Earth hypothesis was that no one could see how a global ice pack could ever have formed in the first

---

[16] There are, however, some other interesting examples from paleobiology that do not involve simulation of macroevolutionary processes. See, for instance, Hughes (1999), who describes the use of computer imaging techniques to study deformities in fossils.

place. In the 1960s, scientists knew that the much more recent Pleistocene glaciers had not extended all the way to the tropics. This situation changed when M. I. Budyko developed a climate model that incorporated what is known as the ice-albedo feedback effect. Albedo is a measure of the degree to which solar energy is reflected by ice and snow. Budyko's climate model demonstrated that as the area of the earth's surface covered by ice increased, the albedo effect would increase, too, thus resulting in a further cooling of the earth's climate, which in turn would cause the ice pack to grow. Budyko's simulations showed that beyond a certain point, this ice-albedo feedback effect would lead to runaway glaciation. No one took the results of these early numerical experiments to be a confirmation of the snowball Earth hypothesis; instead, the results were rightly taken as a mere proof of possibility. Once the climate models showed how a snowball Earth episode could have gotten started, it became easier to take the snowball Earth hypothesis seriously as a potential explanation of the neoproterozoic glacial deposits.

Over the last few years, geologists and paleoclimatologists have run more and more sophisticated numerical experiments. Using climate simulations, it is actually possible to rewind the tape of geological processes (so to speak) and then play it back, holding some parameters constant while varying others. In other words, computing power makes it possible to do virtually what we earlier imagined a super-experimenter doing actually. Although this is not an example of manipulation of past events and processes, or even of testing novel predictions derived from claims about those past events and processes, it is possible to test predictions about what will happen when we run the simulation while varying certain initial conditions.

The recent numerical experiments have yielded some interesting results. For example, Hyde et al. (2000) developed a numerical model that combines climate information with information about the structure of the ice sheets. They ran several experiments with this model to see (among other things) what would happen if they varied the amount of carbon dioxide in the atmosphere. This is important, since the geological record does not give us any clear evidence as to the amount of carbon dioxide that existed in the neoproterozoic atmosphere. When they ran simulations based on a higher initial amount of carbon dioxide, and when they used an additional model that takes into account the patterns of circulation of heat in the earth's oceans, they found that instead of a snowball Earth episode they ended up with what others have come to call a "slush-ball Earth," or a situation in which continents are covered with glaciers and the seas are mostly frozen over, but with a band of open ocean in the tropics. Here again, the paleoclimate model has served as an important proof of possibility by showing that a runaway ice-albedo effect does not necessarily have to lead to a full-blown snowball Earth episode.[17]

However, as Huss also points out in his discussion of numerical experimentation in paleobiology, this type of experimentation cannot actually solve the basic problem of empirical testing. What Hyde et al. have shown is that some initial conditions would have led to a snowball Earth episode, while others could have led to a near-snowball Earth episode with some open water remaining in the tropics. But unfortunately we do not know which set of initial conditions actually obtained during the neoproterozoic. This leaves us with two distinct and incompatible models, and with the problem of finding some sort of empirical test that could discriminate between them. In conclusion, numerical experimentation is best understood as one of several strategies that historical scientists have developed for coping with the asymmetry of manipulability. But numerical experimentation can only take us so far; ultimately there is no substitute for intervention in nature.

---

[17] In another related case, Donnadieu et al. (2003) used a numerical ice sheet model to disarm one potential objection to the snowball Earth hypothesis. The objection is that the glacial deposits found in neoproterozoic strata could only have been formed by dynamic terrestrial glaciers – glaciers that expand and recede, scraping the earth and depositing rock as they go. How could a global ice pack have the same effects? Donnadieu et al. used a numerical model to show that some precipitation would continue even during a global snowball Earth episode, making for dynamic glaciers. Thus, numerical models can help demonstrate that a hypothesis really can explain the phenomena.

6

Making prehistory: could the past be socially constructed?

The dinosaur is the totem animal of modernity. By this I mean, first, that it is a symbolic animal that comes into existence for the first time in the modern era . . .
W. J. T. Mitchell, *The Last Dinosaur Book* (1998, p. 77).

Conceptual idealism is a ludicrous and anti-scientific view of the world. Science teaches that the moon existed long before we or any other concept-mongers did and is not the sort of thing that can be created by thought or talk.
Alan Musgrave (1999, p. 351).

Two positions: realism and nonrealism. And essentially we seem to be no further ahead than when we started!
Michael Ruse (1999, p. 253).

Did paleontologists discover dinosaurs or did they make them? Is it even conceivable that the past could depend on the present? I will argue, contra scientific realists, that for all we know the distant past could be socially constructed, but that there is also no good reason for thinking that it is socially constructed. Like the religious skeptic who admits that the stories of the Christian tradition might be true, but sees no good reason for thinking that they are true, I counsel agnosticism on the whole question of metaphysical realism *vs.* social constructivism. Someone who follows this policy will endorse neither the radical constructivist view of W. J. T. Mitchell, nor the realist metaphysics of Alan Musgrave.

6.1 WHAT DOES IT MEAN TO SAY THAT SOMETHING IS SOCIALLY CONSTRUCTED?

To assert that something is "constructed" is to assert that it is mind-dependent. Mind-dependence can be understood in different ways, but all of those ways involve the use of counterfactuals. It may seem unfair both to metaphysical realists and constructivists to portray them as being committed to a host of counterfactual claims. Counterfactuals raise notoriously difficult philosophical problems, and charity might require us to explore ways of explicating claims about mind-dependence and -independence without resorting to them.[1] One possible alternative would be to analyze mind-dependence claims as causal claims. On this construal, the claim that dinosaurs are social constructs would amount to the claim that scientists (somehow) caused dinosaurs to exist. Not only is this causal claim obviously false, but if social constructivism is to make any sense at all, we need to interpret the constructivist as being committed to the same sorts of causal claims that scientists are committed to. I will proceed here on the assumption that metaphysical realists and constructivists are indeed committed to counterfactuals. If there were some other way of explicating claims about mind-independence and -dependence without helping oneself to counterfactuals, then the arguments of this chapter might not apply to metaphysical realism and constructivism so construed.

One word of terminological clarification is in order. Realists who reject social constructivism usually do so by affirming that things posited by scientists are mind- and theory-independent. The sense of "independence" that is relevant here is metaphysical independence (a notion that still needs some clarification). This is quite different from the notion of independence that we encountered back in chapter 5. There we saw how some realists, such as Leplin, place an independence condition on predictive novelty. The independence which matters in that context, however, is epistemic independence.

---

[1] An anonymous reviewer suggested that claims about mind-dependence and mind-independence might imply counterfactual claims, without themselves being counterfactual claims. I do not see how this solves the problem, however. If the mind-dependence and -independence claims imply counterfactuals, then philosophers who affirm the former will still be committed to the latter.

Consider the following two claims:

(1) *x* would never have existed if no minds had ever existed.
(2) *x* would not be an *F* if no minds had ever existed.

Claim (1) says that *x* exists mind-dependently, while claim (2) says that *x* has a certain property mind-dependently. Plenty of things are mind-dependent in both of these senses. Take, for example, the Brooklyn Bridge. It seems true that the Brooklyn Bridge never would have existed if no minds had ever existed. So the existence of the Brooklyn Bridge is (in a sense) mind-dependent. On the other hand, consider a rock that someone picks up and uses as a paperweight. That rock would have existed even if no minds had ever existed, but it would not be a paperweight if no minds had ever existed. So although the existence of the rock is mind-independent, it has at least one property mind-dependently. This means that when someone says that something is "constructed," we should immediately ask whether the person is asserting that something exists mind-dependently, or whether it has a certain property mind-dependently.

Common sense tells us that a great many things, especially artifacts, are constructed in either or both of these senses. Trivially, many things have some properties mind-dependently. For example, suppose that "*F*" stands for the property of being thought of by me. Then it is true that Mt. Everest would not be an *F* if no minds had ever existed. But it is also easy to see how philosophical controversies can arise over claims having the form of (1) or (2). Every so often in the history of philosophy an idealist has come along and argued that just about everything is constructed. For example, Bishop Berkeley held that ordinary physical objects such as rocks, stars, trees, and even our own bodies, are constructed in the sense of (1). For he held that all of these things are ideas existing only in minds.

Kant, who strenuously argued against Berkeley's view that all physical objects exist mind-dependently, nevertheless held that just about everything is constructed in the sense of (2). Suppose, for example, that a tree has a spatial property, such as the property of being twenty-five feet tall. Kant's view was that the tree – or something, the *Ding an sich* – would exist even if no minds had ever existed, but that the tree would not have any spatial properties if no minds had ever existed. Interestingly, both Berkeley and Kant were constructivists about observables. That is, they both held that all the objects of ordinary experience are, in some sense, constructed by minds. In philosophy of science today, the big question is whether *unobservables* are constructed.

Notice that it is possible to generate different kinds of mind-dependence claims by tweaking the counterfactuals (1) and (2):

(3) x would immediately cease to exist if all minds were suddenly annihilated.

(4) x would immediately cease to be an F if all minds were suddenly annihilated.

Notice that although the Brooklyn Bridge is mind-dependent in the sense of (1), it is not mind-dependent in the sense of (3). If all minds were suddenly annihilated, the Brooklyn Bridge would presumably continue to exist, at least for awhile, until it fell into disrepair and eventually collapsed. Suppose, on the other hand, that some people build a castle on top of a hill. We can then say that the hill has the property of supporting a castle. It has that property mind-dependently in the sense of (2), because it would never have supported the castle if no minds had ever existed. But the property is not mind-dependent in the sense of (4), for if all minds were suddenly annihilated, the hill would continue to support the castle. Berkeley and Kant held that ordinary physical objects are constructed in the sense of (3) and (4) respectively.

Hopefully it is becoming clear that many claims to the effect that something is or is not constructed are highly ambiguous. Moreover, some claims about construction are just obviously true, while others, such as the claims of Berkeley and Kant, may seem wildly implausible at first, although they may begin to seem more plausible the more one thinks about them.

What does it mean to say that something is socially constructed? So far, I have tried to explicate the notion of construction by talking generically about dependence or independence with respect to minds, but whose minds, exactly, do I have in mind? Probably neither Berkeley nor Kant deserves to be called social constructivist, because both philosophers tended to think of construction as dependence on a single mind. To say that something is socially constructed, then, is to say that something depends for its existence and/or its nature upon the thoughts, views, desires, intentions, opinions, assumptions, etc. of some specified community. This introduces another ambiguity into claims about social construction, because it is not always clear just which community we are talking about. One interesting kind of social constructivism can be generated by formulating mind-dependence claims with respect to the scientific community:

(5) *x* would not exist if the scientific community thought differently.

(6) *x* would not be an *F* if the scientific community thought differently.

One potential problem is that members of the scientific community do not always agree about everything. There are, however, many things that they do agree upon. For example, virtually everyone in the scientific community thinks that oxygen exists, and that it has an atomic number of 13. A social constructivist might assert that oxygen would not exist at all if the scientific community thought differently, or that oxygen would not have the atomic number 13 if the scientific community thought differently. Such claims may strike us as a bit odd, but they are no more bizarre than the views of such unquestionably great philosophers as Berkeley and Kant. This sort of constructivism is sometimes advocated by sociologists and anthropologists of science, but seldom by philosophers of science. At any rate, social constructivists generally want to defend claims such as (5) and/or (6).

In his recent book, *The Social Construction of What?* (1999), Ian Hacking highlights another ambiguity in many social constructivist views. It is often unclear just what sorts of things are supposed to be socially constructed. Earlier on, we saw that claim (1) asserts that something exists mind-dependently, while (2) asserts that something has a property mind-dependently. Instead of focusing on things and their properties, we could just as easily focus on events, processes, or facts.

Can the past be constructed? This sort of question is most frequently raised by people interested in what is sometimes called "revisionist history," and sometimes also by those interested in psychotherapy. Consider, for example, the following bizarre claims:

(7) The Holocaust would not have occurred if the community of historians did not now think that it occurred.

(8) Jones would not have been molested as a child if he did not now think he was molested.

(9) The dinosaurs would have been cold-blooded if the scientific community did not now think that they were warm-blooded.

These sorts of claims are bound to strike one as eccentric at best, insidious at worst. Although (7) and (9) are claims of social construction, while (8) concerns only individual construction, all assert that the past depends, in some sense, on the thoughts that people have now, in the present. How can that be? How could the past possibly depend on the present? Is that not absurd?

6.2 FIVE ROADS TO SOCIAL CONSTRUCTIVISM, ALL PAVED WITH GOOD INTENTIONS

Perhaps the first question that needs to be addressed is why anyone would ever *want* to be a social constructivist, except of course about uncontroversial things like artifacts. Here it is interesting to compare social constructivism with other radical philosophical views, such as pyrrhonian skepticism and external world skepticism. Self-professed skeptics are hard to find in the history of philosophy, and yet plenty of philosophers have proudly endorsed idealism (Berkeley, Kant), and recently social constructivism has become quite popular in some academic circles, if not among professional philosophers of science. Social constructivism about the past is such a weird view, however, that one would need to have pretty strong philosophical motivations for adopting it. What could those motivations be? I will focus here on the motivations for going constructivist rather than on the arguments for constructivism, because I doubt that anyone since Kant has actually given any terribly good arguments for such a view. There are at least five distinct motivations for going constructivist: first, the desire to avoid skepticism; second, the desire to correct the excesses of various kinds of realism and positivism; third, the desire to show that some politically distasteful or objectionable situation could be different from what it is; fourth the desire to embrace the consequences of certain theories of truth for which there are independent motivations; and fifth, the desire to provide the most accurate possible description of scientific practice.

Anti-skeptical motives. This is clearly what motivates the philosophies of Berkeley and Kant. Berkeley, for instance, held that materialism (the belief in the existence of mind-independent material stuff) leads inexorably to skepticism. All of our knowledge is based on experience, but we have experience only of our own ideas. Therefore, the way to guarantee the possibility of empirical knowledge, and to vanquish Cartesian external world skepticism, is to assert that physical objects just *are* ideas – that is, that their existence is mind-dependent. Kant was also motivated to go constructivist by his desire to avoid Humean skepticism about induction. In light of this tradition, it would not be surprising if some philosophers embraced constructivism about unobservables in order to avoid skepticism. For example, constructivism might be one way to avoid the skeptical consequences of the global underdetermination argument or the pessimistic induction. The constructivist gambit is to guarantee the possibility of knowledge by making the object(s) of knowledge depend, in one way or another, on the knowing subject.

Anti-realist motives. Much of the social constructivism of the late twentieth century takes its inspiration from the work of Thomas Kuhn, who in turn takes his inspiration from Kant. In his classic, *The structure of scientific revolutions*, Kuhn makes some tantalizing and controversial remarks about scientific change, suggesting that the world itself changes when scientists replace old paradigms with new ones:

> On at least seventeen different occasions between 1690 and 1781, a number of astronomers, including several of Europe's most eminent observers, had seen a star in positions that we now suppose must have been occupied by Uranus. One of the best observers in this group had actually seen the star on four successive nights in 1769 without noting the motion that could have suggested another identification. Herschel, when he first observed the same object twelve years later, did so with a much improved telescope of his own manufacture. As a result, he was able to notice an apparent disk-size that was at least unusual for stars. Something was awry, and he therefore postponed identification pending further scrutiny. That scrutiny disclosed Uranus' motion among the stars, and Herschel therefore announced that he had seen a new comet! Only several months later, after fruitless attempts to fit the observed motion to a cometary orbit, did Lexell suggest that the orbit was probably planetary. When that suggestion was accepted, there were several fewer stars and one more planet in the world of the professional astronomer. (Kuhn 1996, p. 115)

It is by no means easy to know just how to interpret this passage and others like it. Kuhn himself may not have wanted to defend social constructivism at all, and there are surely ways of interpreting the phrase "in the world of the professional astronomer" that would be compatible with realism. Yet if we take that last sentence at face value, it looks like Kuhn is saying that the number of planets is mind-dependent. Before the discovery of Uranus, there were seven planets; afterwards, there were eight. Kuhn, in other words, appears to be making a broadly Kantian constructivist claim along the lines of (6) above. Uranus is a planet. However, Uranus would not be a planet (but rather a star or a comet) if the scientific community thought differently. Perhaps the most charitable interpretation is to suppose he did not intend for that sentence to be taken literally, because it clearly does not follow from the historical claims that precede it. A scientific realist could agree with Kuhn about all the historical facts, and then insist that there were nine planets all along (if not more) and that earlier scientists were simply mistaken in thinking that there were only seven (or eight). What would lead Kuhn to say such things as that "after a [scientific] revolution scientists are responding to a different world" (1996, p. 111)? Certainly no facts about the history of science imply this sort of claim.

One possible suggestion is that Kuhn's motivation for going constructivist is mainly polemical. First, imagine a political debate in which a centrist candidate is attacking the position of a more conservative candidate with respect to a particular issue. In such contexts, there is always a motivation to distance oneself as much as possible from the position one is criticizing. As the polemic heats up, the centrist candidate may drift further to the left, so as to disassociate herself from the conservative target of her arguments. It is not unreasonable to suppose that something similar is occurring in Kuhn's case. Second, Kuhn's main goal is to show that there is nothing about the history of science that requires a realist interpretation. Nothing requires us to say that later theories are more approximately true, or likelier to be true, than the theories they have replaced. Nor are there any particularly good reasons to suppose that unobservables exist and have the properties they do independently of our theories and paradigms. It could be that Kuhn's constructivist claims serve mainly to accentuate these points by showing that a pretty radical anti-realist (that is, constructivist) interpretation is perfectly compatible with the history of science. Perhaps Kuhn is trying to show that realism is optional by showing that constructivism is an option, too.

Even if this is not what Kuhn was up to, the point is simply that a philosopher could be motivated to go constructivist for polemical reasons – whether to distance oneself from one's opponents, or to show that view diametrically opposed to the opponents' view is still coherent.

Political Motives. In addition to anti-realist polemics, Hacking (1999, p. 6) points out that many social constructivists are engaged in a different sort of polemic. In many cases, the point of claiming that X is socially constructed is to say that "X need not have existed, or need not be at all as it is." Social constructivists who make such claims often also want to say that things would be better if X did not exist, or if X were different. For example, someone who says that race is a social construct may well have a political motive for making that claim. He may want to assert that it would be better if there were no such thing as race, or if common sense racial categorizations were different.[2] No doubt constructivists

---

[2] For a good example of a philosopher of science who argues that race is a social construct, and who may have political motivations for doing so, see Zack (2002).

about human history also have political motives for thinking what they do.[3]

The route to constructivism via theories of truth. Another prominent philosopher who has flirted with constructivist views about the past is Michael Dummett (see, e.g., his 2003). Since my main concern in this book is not with philosophy of language or with theories of truth, I will not try to sort through the details of Dummett's work here. In particular, I shall pass over Dummett's arguments for a justificationist theory of meaning. Suffice it to say that Dummett is attracted to justificationist theories of meaning and truth, but he is aware that such theories have strange implications when applied to statements about the past. We can see what the problem is by considering an oversimplified justificationist (or epistemic, or warranted assertibility, or verificationist) theory of truth for statements about the past. Let "P" stand for any statement about the past.

> "*P*" is true if and only if the available evidence is sufficient to justify asserting that *P*.

In other words, we may define truth (for statements about the past) in terms of justification, or warranted assertibility. This theory has some bizarre implications, however. Consider, for example the statement, "*Pachycephalosaurus* had a neon blue patch on its head." This statement cannot be true on the present theory, because the available evidence does not justify asserting it. On the other hand, the available evidence does not justify our asserting that *Pachycephalosaurus* did not have a neon blue patch on its head, either. So is this statement neither true nor false? Notice that this is the problem of local underdetermination rearing its ugly head again: Any attempt to define truth in terms of justification or warrant will yield the result that statements that are underdetermined by the evidence are neither true nor false, in violation of the law of bivalence. Dummett sees this problem clearly:

> If that [justificationist] account of meaning demanded that we allow as true only those statements about the past supported by present memories and present evidence, then large tracts of the past would continually vanish as all traces of them dissipate. (Dummett 2003, p. 28)

---

[3] For a vivid example of politically motivated constructivism in archaeology, see Gero (1989).

As the traces dissipate – or as historical processes destroy information about the past – statements about the past would lose their truth values. This runs contrary to our pre-theoretical intuitions about truth:

> We could not so much as think of a statement about the past as having once been true, though now devoid of truth value, save in terms of present evidence that evidence for its truth once existed. This conception, though not incoherent, is repugnant: We cannot lightly shake off the conviction that what makes a statement about the past true, if it is true, is independent of whether there is *now* any ground that we have or could discover for asserting it. (2003, p. 28)

The difficulty of shaking this realist conviction is going to be a problem for any warranted assertibility account of truth, not just for such an account of truth with respect to statements about the past. At any rate, Dummett suggests that we could rescue this intuition in the following way:

> But if the truth of a proposition consists of its being the case that someone suitably placed *could have* verified it, or have found a cogent ground for asserting it, then our conviction is vindicated. (2003, p. 28)

This is akin to saying that dinosaurs are observable, because we could have observed them if only we had been "suitably placed" – i.e. if only we had been there seventy million years ago. This philosophical move leads to a rather strange modification of the theory of truth sketched above:

> "*P*" is true if and only if the evidence available to someone "suitably placed" would have been sufficient to justify asserting *P*.

One problem with this move, according to Dummett, is that this leaves it a mystery concerning how people come to understand statements about the past. The whole point of justificationist semantics, to begin with, was to explain how people come to understand statements about the past, and it accomplished this by explaining the meaning and truth of those statements in terms of the evidence that would be sufficient to justify asserting them.

The problem we have been considering is not obviously related to constructivism. The problem is simply that a straightforward justificationist theory of truth will yield the result that a vast number of statements about the past have no truth values at all, or as Dummett puts it elsewhere, that there are "gaps in reality: questions to which there is no answer one way or the other" (2003, p. 40). According to this straightforward justificationist theory, there simply are no facts about the colors of the dinosaurs. Facts about the past "continually vanish" as historical processes destroy the evidence. But as counterintuitive as this is, it seems to fall short of constructivism. A constructivist would say that facts about the past depend counterfactually on the thoughts, theories and paradigms of the present scientific community, or perhaps on the evidence available to the present community.

The straightforward justificationist theory of truth for statements about the past does imply that facts about the past are mind-dependent. In order to see why, consider the following instance of Tarski's T-schema:

"Uranus is a planet" is true if and only if Uranus is a planet.

Notice that according to the unmodified justificationist theory, whether "Uranus is a planet" is true depends on whether the evidence justifies asserting that Uranus is a planet. But then this instance of the T-schema implies that whether Uranus is a planet also depends on whether the evidence available justifies asserting that Uranus is a planet. Here is the argument laid out more explicitly.

P1.   "P" is true if and only if P.
P2.   "P" is true if and only if the presently available evidence is sufficient to justify asserting that P.
C.   P if and only if the presently available evidence is sufficient to justify asserting that P.

When the justificationist theory of truth (P2) is combined with the T-schema (P1), we get the result that Uranus is a planet if and only if the presently available evidence is sufficient to justify asserting that Uranus is a planet. However, if the available evidence were different – if, say, the evidence were sufficient to justify asserting that Uranus is a star – then we could infer, using the T-schema, that Uranus is a star. Oddly enough, the justificationist theory of truth leads toward the position that Kuhn flirted with. In order to see how, just imagine that at one point in time, t_1, the evidence is sufficient to justify asserting that a certain object is a star. At a later time, t_2, the evidence is sufficient to justify asserting that the same object is a comet. At a still later time, t_3, the evidence justifies asserting that the object is a planet. Taken together, the T-schema and the justificationist theory of truth imply that the world itself changes in lockstep with changes in what scientists are justified in asserting.

This is a somewhat more complicated variety of constructivism than the fairly simplistic varieties I considered at the beginning. When applied to history, however, the justificationist theory of truth yields the result that the past depends counterfactually on the present – or more precisely, that the past depends counterfactually on what people are now justified in asserting. This will lend no aid and comfort to Holocaust deniers and other oddball revisionist historians, because no one is justified in asserting that the Holocaust never occurred. Yet suppose that at time t_1 scientists are justified in asserting that dinosaurs were cold-blooded. At some later time, t_2, they have more evidence – more fossils, better techniques for studying bone microstructure, and so on. Now, at t_2, scientists are justified in asserting that some dinosaurs were warm-blooded, like birds. The justificationist theory of truth, together with the T-schema, implies that a fact about the past (i.e. whether dinosaurs were warm- or cold-blooded) changes between t_1 and t_2. As strange as it sounds, whether dinosaurs were warm- or cold-blooded depends counterfactually on what scientists are now justified in asserting.

Thus, one road leads to constructivism via the philosophy of language and, as we have seen, via an epistemic (a.k.a. justificationist, verificationist, warranted assertibility) conception of the nature of truth. What would motivate someone to defend such a theory of truth? Interestingly, one motivation could be the desire to avoid radical skepticism. The justificationist theory of truth effectively eliminates the possibility that I am justified in asserting that I have two hands and yet do not in fact have two hands, because I am a brain in a vat. Another motivation might be polemical. Perhaps some philosophers have embraced justificationism in order to show that a realist conception of truth (such as the traditional correspondence theory) is merely optional. Dummett, however, seems to be led to justificationism by independent considerations in the philosophy of language. He seems to think that some such justificationist theory offers the best account of how speakers acquire an understanding of the meanings of sentences.

The phenomenological route to constructivism. I have saved for last the most prominent defenders of social constructivism, Bruno Latour and Steve Woolgar (1986). Latour and Woolgar explicitly claim that scientists construct facts about the world in the course of their laboratory work, and they are not overly concerned with the traditional philosophical problems of skepticism and the nature of truth. Like Kuhn's, their work is partly polemical. However, they seem to think that if one enters the laboratory as an anthropologist and attempts to provide the most accurate description of the phenomena one observes there, one will end up describing the construction of facts:

> From their initial inception members of the laboratory are unable to determine whether statements are true or false, objective or subjective, highly likely or quite probable. While the agonistic process is raging, modalities are constantly added, dropped, inverted, or modified. Once the statement begins to stabilize, however, an important change takes place. *The statement becomes a split entity.* On the one hand, it is a set of words which represents a statement about an object. On the other hand, it corresponds to an object in itself which takes on a life of its own ... Previously, scientists were dealing with statements. At the point of stabilization, however, there appears to be both objects and statements about these objects. Before long, more and more reality is attributed to the object and less to the statement about the object. Consequently, an inversion takes place: the object becomes the reason why the statement was formulated in the first place. (1986, pp. 176–177, emphasis in the original)

If Latour and Woolgar are right, scientists actually create objects and facts about those objects through the course of their laboratory work. Consider how their description, in this passage, of the dialectic in which scientific statements give rise to objects, might apply to the snowball Earth debate in geology. To begin with, scientists formulate statements, or hypotheses about the distant past, such as the statement that the earth was once covered by snow and ice. As recently as the early 1990s, scientists were unable to determine whether such statements are true or false, or even how likely they are to be true. Over the last decade, however, the "agonistic process" has raged, while defenders of the snowball Earth hypothesis have sought to answer their critics. In the last five years or so, the more modest statement that there were once glaciers at the equator has begun to stabilize, and the low-latitude glaciers have, in a sense, taken on a life of their own. Scientists are now attributing "more and more reality" to the glaciers, to the point where they now sound like realists who would insist that the reason for formulating the hypothesis in the first place was to discover the truth about the glaciers. But once we understand the process by which scientists arrive at this point, we understand that geologists have made the low-latitude glaciers. They have *made prehistory*. Or at least that, presumably, is what Latour and Woolgar would want to say. It seems they are led to constructivism because they think that constructivism affords the most accurate description of scientific practice. I call this the "phenomenological" route to constructivism.

My main complaint about the constructivism of thinkers such as Latour and Woolgar parallels the argument that I will shortly give concerning Devitt's metaphysical realism. To say that scientific objects are constructed is to say that they are (in some sense) mind- or theory-dependent, and to say that something is mind- or theory-dependent is, I have argued, to make a special sort of metaphysical claim having the form of a counterfactual conditional. Why would anyone think that descriptions of episodes in the history of science, or close, careful descriptions of scientific practice in the lab, at conferences, or in the field, require any such metaphysical claims? One can easily offer descriptions of scientific practice that are neutral with regard to realism and constructivism: "As scientists become more and more confident that their theory is well-supported by the evidence, they begin to speak and write as though the entities posited by that theory really exist, and as though the events and processes posited by it really occur. Over time, scientists begin to talk less about the theory and more about those things and events, etc."

To summarize: philosophers have been led into the arms of constructivism by the desire to avoid skepticism; by the desire to distance themselves from realist positions they mean to criticize; by a desire to demonstrate that realist positions are merely optional; by the desire to advance a political agenda; as a result of adopting a certain conception of truth for which there may be other, independent motivations; and/or by the desire to offer the closest, most accurate possible description of scientific practice.

6.3 ARE THERE ANY GOOD ARGUMENTS FOR CONSTRUCTIVISM?

Berkeley thought that he had a formidable *a priori* argument for idealism because he thought he could show that the very concept of matter is self-contradictory. Materialists want to define "matter" as a substance existing independently of the mental, and having such qualities as extension and solidity. Since Berkeley thought he could show that all the qualities that materialist philosophers attributed to matter – all the traditional primary qualities – were in fact sensible qualities, and that sensible qualities exist only insofar as they are perceived by a mind, he thought he could show that there is no such thing as matter. It would be impossible for something existing mind-independently to have sensible qualities. I will leave it to historians of philosophy to evaluate this argument. Do contemporary constructivists have any arguments to offer that even come close to the power of Berkeley's? Kukla (2000, pp. 44–45) reports that there are no serious *a priori* arguments for constructivism in the contemporary literature. It seems that contemporary constructivists have preserved much of the philosophical outlook of Berkeley and especially Kant, while dropping their argumentative strategies.

What about *a posteriori* arguments? One possibility is that there are certain facts about the history or practice of science that are best explained by supposing that unobservable things and events are mind-dependent. (For an exploration of this line of argument, see Kukla 2000, chapter 5). There is some indication that Latour and Woolgar develop this line of argument. It is not clear just how such an argument would go, however. Take, for instance, the earlier claim:

(7) The Holocaust would not have occurred if the community of historians did not now think that it occurred.

In order to mount an *a posteriori* argument for this claim that the Holocaust is socially constructed, the social constructivist would need to show either that there is some phenomenon – say, some fact about the practice of historians – that is better explained by (7) than by any other hypothesis, or else derive some prediction from (7) and test that prediction against the empirical evidence. Perhaps I am missing something, but I cannot think of any empirical evidence that would lend any support to (7), or any evidence that (7) explains. And this point generalizes to other constructivist claims.

Latour and Woolgar, perhaps, come closest to offering a serious *a posteriori* argument for constructivism, yet it is not clear just how their argument is supposed to go. They seem to think that every fact has a history: the development of a fact from a statement or hypothesis is a bit like the development of an oak tree from an acorn. The vast majority of statements never do become scientific facts:

> A laboratory is constantly performing operations on statements; adding modalities, citing, enhancing, diminishing, borrowing, and proposing new combinations . . . Each statement, in turn, provides the focus for similar operations in other laboratories. Thus, members of our laboratory regularly noticed how their own assertions were rejected, borrowed, quoted, ignored, confirmed, or dissolved by others. (Latour and Woolgar 1986, pp. 86–87)

Occasionally, out of this interaction, a statement will somehow develop into a fact:

[I]n situations where a statement is quickly borrowed, used and reused, there quickly comes a stage where it is no longer contested. Amidst the general Brownian motion, a fact has been constituted. This is a comparatively rare event, but when it occurs, a statement becomes incorporated in the stock of taken-for-granted features which have silently disappeared from the conscious concerns of daily scientific activity. (Latour and Woolgar 1986, p. 87)

At times, it looks like Latour and Woolgar wish to defend constructivism on the grounds that it affords the best explanation of certain phenomena they noticed in the lab. Why a constructivist explanation is required is not entirely clear. At other times, it seems (as noted earlier) that they wish to say that we can only describe those phenomena accurately in constructivist terms. Yet in the passage just quoted, it seems that what is really going on is that they have an idiosyncratic definition of "fact". They seem to define a fact as any statement that is no longer contested by anyone in the scientific community. If we define facts in that way, then of course Latour and Woolgar are right to think that facts are (in a sense) constructed, that every scientific fact has a developmental history, and so on. What is not clear, however, is why we should conceive of facts in quite the way that they recommend. Intuitively, most people would agree that there is a fact about whether an ant crawled across my desk at midnight last night. There could not possibly be any such fact, in the sense of "fact" recommended by Latour and Woolgar. In order to mount an *a posteriori* argument for constructivism, they would need to give some empirical evidence to support their view concerning the nature of facts. One wonders what sort of evidence could count either for or against a definition of "fact".

6.4 WHY THE ABDUCTIVE ARGUMENT FOR REALISM DOES NOT SUPPORT METAPHYSICAL REALISM

Perhaps more than any other prominent realist philosopher, Michael Devitt has stressed that realism is a metaphysical view. He defines realism as the view that "tokens of most current common-sense and scientific physical types exist independently of the mental" (1991, p. 23). The phrase "independently of the mental" is what Devitt calls the "independence dimension" of realism. According to Devitt, the basic difference between scientific realism and common-sense realism is that the latter affirms the mind-independent existence of observables, while the former affirms the mind-independent existence of unobservables. Devitt makes several important claims about realism. First, he insists that scientific realism is an empirical hypothesis (1991, p. 109). Second, he stresses that the realist "is not committed to all the unobservables of modern science," but only most of them (1991, p. 109). The reason for this caution is that the history of science gives us reason to wonder whether some of the unobservables of current science might later turn out never to have existed. Thus, realism involves a cautious commitment to the ontology of current science. Finally, the realist does not just affirm the existence of things like electrons, muons, gluons, bosons, and all the rest, because realism also includes the independence dimension: "For scientific realism, apart from its independence dimension, *is* a more cautious restatement of the ontological claims about unobservables made by our theories" (1991, p. 114). Devitt, I will show, is mistaken to think that metaphysical realism is an empirical hypothesis.

Although he also endorses the traditional inference to realism as the best explanation of the empirical success of science, Devitt, as we saw in chapter 3, places more stock in what he calls the basic abductive argument for realism. Consider again, by way of illustration, the controversial snowball Earth hypothesis. Hoffman and Schrag (2000; Hoffman et al. 1998), argue that by supposing that the earth underwent one or more episodes of global glaciation at the end of the neoproterozoic era, approximately 800 to 580 million years ago, one can explain (1) puzzling neoproterozoic glacial debris that appears to have been deposited at or near the equator; (2) unusual iron-rich deposits mixed in with the glacial debris; (3) thick layers of carbonate rock that sit right above the glacial debris in places such as Namibia's Skeleton Coast; (4) unusual ratios of carbon 12 to carbon 13 isotopes in neoproterozoic rocks; and (5) the geologically sudden Cambrian explosion that began around 570–560 million years ago. The proponents of the snowball Earth hypothesis argue that we should accept it on the grounds that it affords the best explanatory unification of these observable traces. Obviously, the hypothesized snowball Earth event is unobservable, barring time travel. Devitt's view is that the explanatory power of the snowball Earth hypotheses affords a good reason to think that the entire planet really did freeze over.

The problem with Devitt's view is that while the basic abductive argument does support empirical hypotheses about unobservables, it does not support scientific realism's independence dimension. In order to see why, consider a set of tracks in the snow. What is the best explanation of the existence of tracks having such-and-such features? Suppose we start with the following explanatory hypothesis:

H. A deer passed this way not long ago.

This is just what a realist might say, minus the independence dimension. This conforms with the actual practice of historical scientists. Hoffman and Schrag, for instance, say nothing at all about whether the snowball Earth event occurred mind-dependently or mind-independently. They simply state the hypotheses, all by itself. Devitt, however, wants to add to *H* an independence claim:

H*. A deer passed this way not long ago, *and* the deer's making of the tracks in the snow occurred independently of the mental.

In other words, Devitt conjoins a metaphysical claim to the explanatory hypothesis, and insists that the conjunction affords the best explanation of the observable traces. The question Devitt fails to address is why anyone should think that *H** affords a better explanation of the tracks in the snow than *H* does. This problem is especially acute, because *H* appears to be the better explanation of the two, given any reasonable account of the non-empirical theoretical virtues, such as simplicity and explanatory power. *H* is obviously simpler than *H**, because it involves only one claim as opposed to two. So in order to show that *H** is a better explanation, Devitt needs to show that *H** has more explanatory power than *H*. In other words, Devitt needs to show that the extra conjunct ("and the deer's making of the tracks in the snow occurred independently of the mental") adds something to the explanatory power of *H*. But what does it add?

If Devitt is right to think that the basic abductive argument lends any support at all to *H** as opposed to *H*, that can only be because he has an idiosyncratic conception of explanatory power. In general, what would we need to add to any given hypothesis *H* that potentially explains *E* in order to enhance its explanatory power? There are limited possibilities: (1) We could conjoin some further claim(s) that enhance our understanding of the causal processes leading to *E*. (2) We could conjoin some further claim(s) that widen the scope of *H*, enabling it to explain more than just *E*. (3) We could conjoin some further claim(s) about the processes that led to the processes that led to *E*, thus tracing the explanation further back. (4) We could conjoin some further claim(s) about how *H* is related to other explanatory hypotheses and theories. Unfortunately, the realist's independence claim does none of these things.

We can also approach the issue probabilistically, as follows:

H.   A deer passed this way not long ago.

E.   There are tracks in the snow having such-and-such features.

R.   The deer's making the tracks occurred independently of the mental.

We can say that the prob(E|H) is the *likelihood* of the hypothesis that a deer passed this way not long ago. What Devitt needs to show, in order for the basic abductive argument to succeed, is that prob(E|H and R) > prob (E|H), but he makes no attempt to show this. Nor is it at all clear why conjoining an abstruse metaphysical claim to a hypothesis should enhance the likelihood of the hypothesis. Why would the tracks in the snow be any more probable, given H and R, than given H alone?

I conclude that Devitt has not succeeded in showing that the basic abductive argument lends any support to scientific realism. Indeed, it is not quite true that scientific realism, understood as a claim about the mind-independence of unobservables, is an empirical hypothesis at all. Devitt attempts to pull one over on us by conjoining the realist's independence claim to the empirical hypotheses of science, and then asserting that realism is an empirical hypothesis. Notice, however, that if realism (i.e. the claim that "tokens of most unobservable scientific physical types exist independently of the mental") is an empirical hypothesis then so is constructivism. Devitt makes no attempt to show what sort of empirical evidence would count in favor of the realist hypothesis *H** and against the following constructivist hypothesis:

H**   A deer passed this way not long ago, and the deer's making the tracks was dependent, in one way or another, on our thoughts now.

Tellingly, in the attack on constructivism that Devitt develops in one of the later chapters of his book, *Realism and Truth* (1991), he does not rely on the abductive argument for realism at all, by showing that realism has some non-empirical theoretical or explanatory virtue that constructivism lacks. Nor does he try to show that realism has some empirical virtue that constructivism lacks. Instead, he falls back on a *priori* arguments against constructivism.

How else might Devitt reply to this accusation that the basic abductive argument commits a fallacy of relevance, and that it does not support the independence dimension of realism? One possible move that Devitt could make is to argue that contrary to appearances, *H* and *H** are not really distinct hypotheses at all. He might insist that the mind-independence claim ("The deer's making the tracks occurred independently of the mental") is contained implicitly in *H*, and made explicit in *H**. On the face of it, this is quite plausible. If, every time we made an empirical claim, we took the time to conjoin a mind-independence claim to it, communication would become much more tedious. If I say, "It snowed yesterday," probably most people, if they thought about it, would take this to mean that the snow fell mind-independently. With this in mind, Devitt might deny that *H* is really simpler than *H**, while insisting that *H* derives any explanatory power that it has from the implicit independence claim.

But this reply misses the point. It could be true that when someone says, "A deer passed this way not long ago," that person is usually tacitly making the realist claim that a deer passed this way not long ago, and that event occurred independently of the mental. However, the point at issue is not whether ordinary people tacitly make realist claims, but whether those realist claims are justified. If we want to, then surely we can resolve to examine the claim that the deer passed this way not long ago, all by itself, in order to see how much explanatory power it has. The problem is that we do not enhance the explanatory power of such claims by conjoining them with other metaphysical claims.

Before going on to examine the *a priori* arguments against constructivism, however, it is worth pointing out that exactly the same considerations raised here against Devitt apply with equal force to the constructivist who tries to argue that constructivism can explain some phenomenon better than realism can. The constructivist and the realist have something in common: both want to conjoin metaphysical claims to empirical hypotheses. What I want to suggest, following Arthur Fine, is that neither the realist nor the constructivist has given us any good empirical reason to do that. Both the realist and the constructivist have the same burden: both must show that we can enhance either the empirical virtue or the non-empirical virtue of hypotheses by conjoining them with counterfactual claims to the effect that something is mind-dependent or –independent. No philosopher that I know of has seriously tried to meet this challenge.

6.5 *A PRIORI* ARGUMENTS AGAINST HISTORICAL CONSTRUCTIVISM

Philosophers such as Devitt like to accuse constructivists of making some pretty basic philosophical errors, such as confusing the representation of *x* with *x* itself (see, e.g. Devitt 1991, pp. 239ff.). Constructivists, he says, "blur the crucial distinction between theories of the world and the world itself"

(1991, p. 241). The trouble with this, as Devitt himself acknowledges, is that what looks to one philosopher like a blurring or a confusion can look to another like the deliberate collapsing of a distinction. For example, it would not be quite fair to say that Berkeley blurred the distinction between ideas and physical objects, or that he confused ideas of things in the world with those things. Rather, Berkeley quite deliberately argued that ordinary physical objects just *are* ideas. This collapsing of the distinction between the representation of *x* and *x* itself is, in a sense, the whole point of constructivism. So it will not do simply to assert that constructivists fail to draw the distinction; rather, they draw the distinction and then deliberately collapse it.

Andre Kukla (2000) offers a detailed and rigorous examination of the arguments for and against varieties of social constructivism, including some *a priori* arguments against constructivism. Kukla himself contends that certain strong forms of constructivism lead to "temporal incoherence." I will argue that he confuses incoherence with strangeness, or repugnance. There is nothing absurd or self-contradictory about the view that the past is socially constructed. For all we know, scientists *do* make prehistory. The problem with radical constructivism about the past is not that it is incoherent, but only that no one has offered any good reason to believe it.

Kukla begins with the following argument:

> Suppose that at time t1, we construct the fact X0 that X occurs at an earlier time t0; then, at a later time t2, we construct the fact –X0 that X *doesn't* occur at t0. Then it seems to follow that X0 is true (because that fact was constructed at t1) and that –X0 is true (because *that* fact was constructed at t2). What do constructivists have to say about that? (Kukla 2000, p. 107)

Notice how similar this is to the earlier problems we encountered in connection with the justificationist theory of truth. To make Kukla's argument here more concrete, we can simply rehearse an earlier example: Suppose that at t1 (say, the mid-twentieth century), scientists construct the fact (X0) that all dinosaurs were cold-blooded at an earlier time t0 (say, the Jurassic). Later on, at time t2 (say, the 1980s), scientists construct the fact (– X0) that some dinosaurs were warm-blooded at the earlier time t0 (the Jurassic). It looks for all the world like social constructivism leads to a blatant contradiction: all dinosaurs were cold-blooded (because that fact was constructed at t1) and some dinosaurs were not cold-blooded (because that fact was constructed at t2). This elegant line of argument appears to show that radical constructivism leads to absurdity.

What Kukla fails to appreciate is that when doing armchair metaphysics, it is always possible to avoid inconsistency by making the appropriate maneuvers. In this case, the social constructivist can point out that Kukla is making some undefended assumptions about the nature of facts. For example, he assumes that when fact X0 is constructed at time t1, it remains a fact until t2, so that at t2 there are two contradictory facts: X0 and not X0. That, of course, is an absurd result. However, the constructivist can avoid this result simply by insisting that the facts about the past change over time as scientists change their views. Recall the earlier suggestion, in connection with the justificationist theory of truth, that what is true about the past changes in lockstep with what scientists in the present are justified in asserting. The constructivist can make the same point here. X0 is a fact at t1, while −X0 is a fact at t2. The facts about whether dinosaurs are warm- or cold-blooded have changed from t1 to t2. At no point, however, do we get the absurd conjunctive fact that all dinosaurs were cold-blooded and some dinosaurs were warm-blooded. If facts about the distant past can literally disappear as historical processes destroy information about earlier things and events, as constructivists must maintain, why not also say that the facts about what existed or occurred at t0 can actually change at later times?

No doubt many philosophers will reply that for constructivists, the price of logical consistency is a bizarre, even a repugnant view about facts. How can facts disappear? Or change over time? This is certainly not what most of us have in mind when we ordinarily talk about the facts. Michael Dummett writes that he long ago arrived at the conclusion that "antirealism about the past was not incoherent; but it was not believable either" (2002, p. 29). In metaphysics, strangeness and repugnance are never as bad as absurdity, and one can always avoid absurdity by making one's view stranger and more repugnant. This goes a long way toward explaining why so many philosophers in the empiricist tradition, from Hume to van Fraassen (2002), have found metaphysics itself to be strangely repugnant.

Kukla continues:

> Let X0 be the fact that X occurs at time t0, and −X0 be the fact that X doesn't occur at t0. Also let C(X0) be the fact that X0 is constructed at t1, and let C2(−X0) be the fact that −X0 is constructed at t2. Now the world at t1 has a past that contains the event X0, and the world at t2 has a past that contains the event −X0. But *pastness is transitive*: If event X is in the past relative to event Y, and event Y is past relative to event Z, then X

is past relative to Z. Moreover, X0 is in the past relative to C1(X0), and C1(X0) is in the past relative to C2(–X0). Therefore, by transitivity, X0 is in the past relative to C2(–X0). That is to say, the world we construct at t2 has in its past the fact X0 that was constructed at t1. But it also has the fact –X0 that was constructed at t2. Therefore the attempt at segregating the contradictories fails. (Kukla 2000, p. 108, emphasis in the original)

Let's call this *the transitivity argument,* since the difference between it and the previous argument is that it incorporates as a premise the transitivity of pastness. Kukla thinks that this argument demonstrates the incoherence of constructivism in general, and of constructivism about the past in particular.

It seems plausible to suppose that pastness is transitive: from the fact that the Revolutionary War is in the past relative to the Civil War, and the fact that the Civil War is in the past relative to the Spanish American War, we may infer that the Revolutionary War is in the past relative to the Spanish American War.

Nevertheless, the constructivist can avoid the conclusion of the transitivity argument. In order to make things as concrete as possible, let us suppose that X0 is the fact that all dinosaurs were cold-blooded. C1(X0) is the fact that this fact about dinosaur metabolism was constructed, say, in the 1930s. –X0, as before, is the fact that some dinosaurs were warm-blooded. And we can suppose that C2(–X0) is the fact that –X0 was constructed at a later time, say in the 1980s. Kukla's argument involves the following crucial move:

P1. The fact that the dinosaurs were cold-blooded (X0) is in the past relative to C1(X0).
P2. C1(X0) is in the past relative to C2(–X0).
C.   So, by the transitivity of pastness, the fact that the dinosaurs were cold-blooded is in the past relative to C2(–X0).

Remember, though, that the constructivist thinks that the facts about the past can change over time. According to the constructivist, up until the 1980s it was a fact that all dinosaurs were cold-blooded. However, with the change of scientific opinion during the 1980s, the facts changed, too. With this in mind, the constructivist will want to say that the phrase "in the past relative to" is ambiguous, and consequently that there are two quite different ways of interpreting the above conclusion. According to the first interpretation, the conclusion is true. X0 is in the past relative to C2(–X0), but that is because at t2, when –X0 is constructed, X0 ceases to be a fact at all. Understood in this way, the conclusion is true, but it poses no threat to constructivism. On this first interpretation, to say that one fact is in the past relative to another is to say that that *the first fact was a fact at an earlier time, and that the second fact was a fact at a later time.*

Someone who says that one fact is in the past relative to another could also mean that *the first fact is a fact about an earlier time, while the second fact is a fact about a later time.* If this is what the phrase means, then the constructivist can and should deny the transitivity of pastness in just those cases where the facts about the past change owing to scientific construction. Imagine that we are at t2, in the 1980s. Is X0 in the past relative to us now? Yes and No. X0 is in the past relative to us now, in the sense that it was once, but is no longer, a fact. In another sense, X0 is definitely not in the past relative to us now. For it is not now a fact about an earlier time. From our point of view at t2, we can say that X0 was once in the past relative to C1(X0), and we can say that C1(X0) is in the past relative to C2(−X0). But since X0 is no longer a fact at all, it cannot now be in the past relative to anything, in the second sense of "in the past relative to . . ." So what has Kukla shown? Only that constructivism about the past is weird, not that it is absurd. Anyone who thinks that facts about the past are socially constructed must bite the bullet and say that facts about the past can change. But as is usually, if not always the case in abstruse metaphysics, anyone who wants badly enough to be a constructivist can find a way to do so consistently.

Kukla (2000, pp. 110–112) is well aware of this rejoinder to the transitivity argument. So he has one more go at constructivism. In order to understand his argument, it will be necessary to reproduce a diagram that he uses (see below).

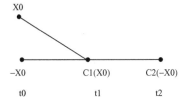

It is important to show the diagram here, because I think it is quite misleading insofar as it fails to show how the facts change over time. At any rate, here is Kukla's final argument against constructivism:

> So there's a history where C1(X0) has X0 in its past (the oblique line), and another history where C1(X0) has −X0 in its past (the horizontal line). But

if C1(X0) can have either X0 or –X0 in its past, then the occurrence of C1(X0) can hardly be said to constitute X0. Evidently, the occurrence of C1(X0) has no bearing on whether X0 occurs in the past. The argument is entirely general – it applies to any putative construction for a past event. (2000, p. 112)

This new argument avoids the earlier problem about there being two different senses in which one fact can be in the past relative to another. The problem here is that the phrase "has no bearing" is ambiguous. Everything here depends on the perspective we choose to adopt. Suppose we adopt the perspective of the later scientists, at t2. From their point of view, the constructive activities of earlier scientists, at t1, once had a bearing on whether dinosaurs were cold- or warm-blooded, but the activities of those earlier scientists no longer have a bearing on that question, because we at t2 have remade prehistory. So if we adopt the perspective of the later scientists at t2, the conclusion that "the occurrence of C1(X0) has no bearing on whether X0 occurs in the past," is correct, but is not contrary to constructivism at all. This is what the constructivist should say if we interpret the phrase "has no bearing" to mean *currently has no bearing*. If, on the other hand, when Kukla uses the phrase "has no bearing" he means *never had any bearing*, the constructivist can and should simply reject the conclusion of the above argument. The constructivist can say, without any absurdity, that for awhile at t1, C1(X0) did have a bearing on whether X0 occurs in the past, but that now, at t2, the constructive activities of those earlier scientists no longer have any bearing on this.

I conclude that Kukla's arguments do not refute social constructivism about the past. For all anyone knows, it could be that we make (and remake) prehistory.

6.6 THE NATURAL HISTORICAL ATTITUDE

I hope that readers will be struck by how little the debate about constructivism actually has to do with science. At any rate, we can now pull all of the results of this chapter together to form an argument:

P1. The *a posteriori* arguments for metaphysical realism about the past do not support any of the realist's claims about the mind-independence of unobservables; nor are there any credible empirical arguments for the constructivist's claims that past unobservables are mind-dependent.

P2. The *a priori* arguments against constructivism do not show that constructivism about the past is incoherent or self-contradictory, and there is no reason to think that the realist's mind-independence claims are absurd or self-contradictory.

C1. Neither the *a posteriori* nor the *a priori* arguments come close to settling the issue whether unobservables are mind-dependent or mind-independent.

C2. Therefore, we should make it a policy to remain agnostic about the realist's mind-independence claims as well as the constructivist's mind-dependence claims.

C1 says that with respect to counterfactual claims about the mind-dependence or –independence of things past, we are in a state of equipollence. The classic example of a state of equipollence is the ancient skeptics' observation that there is no reason whatsoever to believe that the number of stars in the cosmos is either odd or even. Everyone can agree that we should be agnostic with respect to a question like that. I hope to have shown in this chapter that after considering the main arguments for and against realist and constructivist metaphysics with respect to the past, we likewise find ourselves in a state of equipollence. The claims of both the realist and the constructivist are intelligible, and the *a priori* arguments do not show that constructivism about the past is absurd or self-contradictory. Moreover, the empirical arguments are completely irrelevant to the dispute between constructivists and realists. *Pace* Devitt, the abductive argument for realism lends no support at all to realism's independence dimension, and it is not clear that constructivists have succeeded in offering any serious empirical arguments for their view at all. This agnostic attitude is what I will call the *natural historical attitude*.

Interestingly, the argument for C1 can be thought of as an underdetermination argument, of the sort discussed back in chapter 2. In order to see how the underdetermination argument can be brought to bear here, recall the earlier example of the tracks in the snow. The realist and the constructivist can be seen as advocating incompatible explanations of the tracks:

H. A deer passed by not long ago, and the deer's making the tracks occurred independently of the mental.

H*. A deer passed by not long ago, and the deer's making the tracks occurred mind-dependently.

Notice that *H* and *H** are empirically equivalent. The same empirical hypothesis ("a deer passed by not long ago") is conjoined with incompatible metaphysical claims. But since the metaphysical claims do not have any impact at all on the empirical consequence class of the hypothesis, *H* and *H** are empirically equivalent. This means that if we grant the further premise that the only evidence that can count for or against a hypothesis is empirical evidence, it follows that the choice between *H* and *H** is radically underdetermined. There is no reason whatever to think that either hypothesis is any likelier to be true than the other. Hence, we are left in a state of equipollence.

Remember also how the discussion of underdetermination in chapter 2 linked up with the role asymmetry of background theories. There I described a local underdetermination problem as one in which background theories give us reason to think that we will probably never find any evidence that would help us to discriminate between the rival hypotheses. That is certainly the case here, because everything we know about historical events and processes gives us reason to think that information about whether something happened mind-dependently or mind-independently will never get preserved in the historical record. For example, information about whether the animal's walking through the snow occurred independently of us will never be preserved in the tracks the animal leaves. Our best background theories imply that metaphysical information – that is, information about the truth or falsity of certain counterfactual dependence claims – is always destroyed, and never preserved by historical processes. Looked at in this way, the agnosticism that I am advocating is forced upon us by what we know about the formation of historical records – viz. that they do not preserve information about the truth or falsity of the realists' and the constructivists' metaphysical claims. They only preserve information about what existed and occurred; they tell us nothing about whether past things and events were mind-dependent or mind-independent.

Is it really true that the arguments for and against metaphysical constructivism leave us in a state of equipollence? Perhaps not. One might reasonably argue that all of our pre-theoretical intuitions favor metaphysical realism, whereas social constructivism commits one to some highly counterintuitive claims, such as that facts about the past can change with the passing of time. Does this intuitive support mean that realism has more going for it than constructivism does, or that it is better justified? I do not think so, for our intuitions about such matters are highly malleable. Speaking autobiographically, as I have gotten used to the natural historical attitude, my own intuition that things in the past have existed and occurred independently of us has eroded. Nor is it clear why intuitions ought to carry any epistemic weight in this context. An appeal to intuition at this juncture seems inconsistent with realists' own metaphilosophical commitments: they want to see realism itself as a theory about science that is empirically well-supported by the data.

In many cases, a persistent state of equipollence leads to indifference, in the sense of not caring one way or the other. Nobody cares in the slightest whether the number of stars in the cosmos is odd or even, or whether dinosaurs were green or blue. Why not? Because we know that we will always be in a state of equipollence with respect to these claims. My hope is that we will likewise eventually become indifferent with respect to claims about the mind-dependence and/or -independence of unobservables.

Arthur Fine likes to characterize his own natural ontological attitude (or NOA) in terms of what he calls the "core position." He thinks of both metaphysical realism and metaphysical anti-realism as inflationist views: Although both share the core position, both want to add something to it. There has been a good deal of discussion (and some understandable confusion) about how to interpret Fine on this point. I shall interpret him as saying the following: the core position just consists of well-supported claims about unobservables – for example, about electrons. Both the metaphysical realist and the metaphysical antirealist/constructivist believe in electrons, and so accept this core position. But both also want to inflate the core position with metaphysics by conjoining metaphysical independence or dependence claims to the scientific claims about unobservables that constitute that core position. To adopt the natural ontological attitude is to take a "Don't know, don't care" stance toward these metaphysical claims, and to accept the core position all by itself. This strikes me as the best way to interpret Fine, and if it is not what he really holds, then it is nevertheless what (I think) he should say. No doubt some philosophers of science have overlooked this metaphysical neutrality interpretation of the NOA for the following reason. Mainstream philosophers of science seldom treat social constructivism as a serious contender. Like Alan Musgrave, they may be inclined to dismiss it as a "ludicrous and anti-scientific view of the world." I have tried to show that the arguments do not support this sort of exaggerated claim. But notice that if you do not take metaphysical antirealism at all seriously, then when Fine talks about realists and antirealists as inflating the core position, it may not occur to you that metaphysical antirealists are what he has in mind.

Unfortunately, Fine does not always make things easy for those who want to figure out just what the NOA amounts to. For example, when explaining what the scientific realist wants to add to the core position, he does not always make it clear what the addition consists in:

> [W]hen the realist and the anti-realist agree, say, that there really are electrons and that they really carry a negative unit charge and really do have a small mass (of about 9.1 × 10^{-28} grams), what the realist wants to add is the emphasis that all this is really so. "There really are electrons, really!" (Fine 1984, p. 37)

This is a little unfair to the realist, but only just a little. What the realist wants to add – what he means when he stomps his foot and says "Really!" – is that the electrons exist and have the properties they do independently of the mental. The bit about independence is crucial, for the social constructivist could happily say: "There really are electrons, really! And they are social constructs."

One potential objection to the natural historical attitude is that anyone who adopts it must abandon an extremely attractive realist view of science. Arthur Fine captures this view nicely in the following passage:

> For realism, science is *about* something; something *out there*, "external" and (largely) independent of us. The traditional conjunction of externality and independence leads to the realist picture of an objective, external world; what I shall call the *World*. According to realism, science is about *that*. Being about the *World* is what gives significance to science. (Fine 1986, p. 150, italics in the original)

Now compare:

> For realism, the science of prehistory is *about* something; something bygone, "external" and independent of us. The traditional conjunction of externality and independence leads to the realist picture of an objective, external series of historical events; what I shall call the *Past*. According to realism, historical science is about *that*. Being about the *Past* is what gives significance to historical science.

The basic argument for adopting the natural historical attitude is quite simple: the realist picture just sketched involves claims to the effect that past things and events – that is, the *Past* – existed and occurred independently of us, our theories and conceptual schemes. (On the other hand, constructivism and metaphysical antirealism involve metaphysical dependence claims that mirror the realist's independence claims). But when it comes to acquiring knowledge of the past, the only evidence we have to go on consists in observable records, remains, and traces. The problem with the realist picture above is that its central metaphysical claims outrun this evidence. At the end of the day, the realist independence claims and the antirealist dependence claims are empirically equivalent rival interpretations of the historical evidence. As good empiricists, we should simply remain agnostic about such issues, and this agnosticism is the most important aspect of the historical attitude. To adopt the natural historical attitude is to remain open to, but not to affirm, the possibility that prehistory is literally something that we humans have made. What should someone who adopts the natural historical attitude say that the sciences of prehistory are about?

The quick and easy answer to this question – and the official answer of anyone who adopts the natural historical attitude – is simply that the sciences of geology, paleobiology, etc. seek to deliver knowledge of past things and events, leaving aside the whole issue of whether those things and events occurred independently of us. Someone who adopts the natural historical attitude can still say that one of the several aims of historical science is to arrive at the truth about the past, but she must remain neutral as to whether this truth is discovered or made.

A number of Fine's critics have voiced perplexity about the natural ontological attitude. Some (such as Musgrave) fail to see how the NOA differs from scientific realism. Philip Kitcher expresses the worry in the following passage:

> I must confess to finding NOA elusive: in his attacks on realism, Fine seems to become an antirealist, and in his rejection of antirealism, he appears to become a realist. Plainly this assessment is at odds with NOA's professed goal of slipping between two antagonistic positions. (1993, p. 134 n. 11)

Perhaps NOA seems elusive to Kitcher because he is not – at least not at this point in the text – carefully distinguishing between the different dimensions of the realism debate. When we focus on the epistemological dimension of the realism debate, Fine does look like a realist: In the next chapter, I will show that he is willing to join forces with realists against epistemological anti-realists, such as Bas van Fraassen. If I am right, then Fine appears to become a realist because he is one – but only epistemologically. If we focus on the metaphysical dimension, Fine's position is perfectly analogous to that of the agnostic who refuses either to affirm or deny the existence of God. One could say, of the agnostic: "In his attacks on theism, he appears to become an atheist, and in his rejection of atheism, he appears to become a theist." Here the trick is to distinguish between rejecting a claim (in the sense of refusing to go along with it) and denying it (in the sense of asserting that it is false). I want to interpret Fine as rejecting both metaphysical realism and antirealism, while denying neither view, just as the agnostic rejects both theism and atheism while denying neither.

Another useful way to get a fix on Fine's philosophical position is to compare the scientific realism debate to the longstanding philosophical debate concerning the nature of truth. Realists about truth typically suppose that the nature of truth consists in correspondence with reality, and that our beliefs and statements are true or false in virtue of the way the world is. Antirealists (including coherence theorists, warranted assertibility theorists, and others) argue that the nature of truth consists in something other than agreement with or correspondence to reality. Coherentists, for example, hold that what makes a statement or belief true or false is its relationship to other statements or beliefs. Warranted assertibility theorists hold that what makes a statement true or false is whether a person would be warranted in asserting it. There is, however a third position, known as minimalism, or deflationism. According to the minimalist view, both the realists and antirealists are mistaken in assuming that the predicate "is true" refers to some real property of statements and beliefs, a property whose nature then requires elucidation. Minimalists argue that an adequate theory of truth could simply describe the way in which the predicate "is true" functions in our language, and that both the realists and the antirealists are engaging in unnecessary metaphysical speculation. Both parties start with the core position, which is an account of how the truth predicate functions, but they go beyond this and tack on metaphysical claims about the nature or essence of truth. We can think of them as "inflationary" theories of truth, just as Fine refers to realism and antirealism as "inflationary" philosophies of science. Just like the minimalist about truth, the NOAer hopes to get away with doing as little theorizing as possible. Given the recent popularity of deflationist theories of truth (some today would even consider deflationism to be the received view among philosophers who specialize in this area) it is an interesting question why NOA – the analogue of deflationism in the scientific realism debate, has not gained more adherents.

Deflationism about truth is not only analogous to Fine's natural ontological attitude; it appears to be part of the natural ontological attitude. Just what the NOA says about truth is one of the topics of the next chapter.

6.7 TWO PREHISTORIES

To adopt the natural historical attitude is to suspend judgment with respect to the metaphysical claims made by realists as well as constructivists. There is, however, another possible view, inspired by Kant, that I have not yet considered. I will call it the "two prehistories" view.

Suppose we distinguish between (a) prehistory as it unfolded independently of us, our thoughts, theories, paradigms, and conceptual schemes, and (b) prehistory as constructed by us. We could call the former independent prehistory, and the latter constructed prehistory, and thereby synthesize the realist and the antirealist metaphysics. Furthermore, we can say that all knowledge of prehistory is restricted to that which the scientific community has constructed. Scientists can only know about prehistory insofar as they have made it. They can know nothing at all about independent prehistory, which must remain forever inaccessible. This "two prehistories" view conflicts with standard scientific realism, for most realists (i.e. those who combine metaphysical with epistemic realism) will want to say that scientists can acquire knowledge of independent prehistory. It also conflicts with the views of many social constructivists who, like Kant's idealist successors, will just deny that there is any independent prehistory at all. What advantages, if any, does the natural historical attitude have over the "two prehistories" view?

The "two prehistories" view is a lovely interpretation of science.[4] It is alluring, because it enables one to make exciting, radical-sounding claims and then in the next breath qualify them so that they do not seem unreasonable. Someone who takes this view can cheerfully say that dinosaurs, evolution, the ice ages, and all the rest are social constructs! But – not to worry – there is still a real, independent prehistory that we have not constructed. One can cheerfully say that we can know nothing at all about the past! But – not to worry – we can still know a great deal about prehistory insofar as scientists have constructed it.

This "two prehistories" view probably deserves more discussion than I can give it here. The main problem with it is its metaphysical excess. To make this clear, consider a simple claim about prehistory, such as that glaciers once covered New England. The realist will say that the glaciers

---

[4] I thank an anonymous reviewer for inviting me to think about this possibility. For a while during the early stages of the project, I myself gravitated toward the "two prehistories" view.

did so mind-independently. The social constructivist will counter that the glaciers covered New England mind-independently. The "two prehistories" view is even more extravagant: According to that view, there is a mind-independent fact about whether the glaciers ever covered New England, but we cannot know what that fact is. On the other hand, we can know that the glaciers as constructed by the scientific community did cover New England. From the point of view of the natural historical attitude, the historical evidence simply does not support any of this philosophical theory; the more elaborate the metaphysical theory gets, the more it outruns the historical evidence.

7

The natural historical attitude

What can we do but disdain the fake modesty: "I don't really believe in trilobites; it is just that I structure all my thoughts about the fossil record by accepting that they existed"?

Simon Blackburn (2005, p. 195)

Perhaps most people just assume, without thinking much about it, that whatever happened in the past must have happened independently of our thoughts, theories, and conceptual schemes – and independently of what scientists tell us about prehistory. Indeed, it may seem natural to assume that prehistory is not something scientists make. I have tried to show that this assumption, however natural it may seem, is indefensible. There are no good arguments for it at all: no good *a priori* arguments, because (contrary to what some philosophers have said) there is nothing absurd or crazy about the idea that the past depends on us. And there are no good *a posteriori* arguments, because the only arguments that realists have produced (for example, Devitt's basic abductive argument) are quite irrelevant to the question whether anything exists or occurs mind-independently.

In philosophy of science today, the leading alternative to realism is the philosophical view that Bas van Fraassen has called *constructive empiricism*. In this chapter, I present van Fraassen's view and show how that view might apply to historical science. My aim in doing so is to use constructive empiricism as a foil for helping to understand the natural historical attitude. The fact that constructive empiricism commits one to some strange views when applied to prehistory – including, I will argue, radical skepticism about the past – is another reason for thinking that the scientific realism debate has been skewed by the failure to consider historical examples. If contributors to the realism debate had looked seriously at paleobiology

163

and geology, they might never have found constructive empiricism to have much appeal as an overarching interpretation of science.

7.1 ARTHUR FINE'S TRUST IN SCIENCE

Thus far, I have characterized the NOA (and also the natural historical attitude) in a negative way, as an agnostic attitude. To adopt the natural historical attitude is to treat metaphysical claims concerning the mind-dependence and/or –independence of the past with skeptical indifference. The overall view of historical science that I have been developing in this book is skeptical in another way, too: All along I have emphasized the limits to our knowledge of the past, and I have argued that the asymmetry of manipulability and the role asymmetry of background theories place historical investigators at a relative epistemic disadvantage, though one that can sometimes be overcome with the help of technology. Indeed, Fine himself often writes as though the NOA were a purely negative, or nay-saying attitude, rather like deflationary theories of truth. Where realists and constructivists think that there is an interesting and important metaphysical issue to be settled by examining the arguments, the NOAer holds that there is no issue at all aside from the problems that scientists are already working on. Thus, Fine tells us that "perhaps the greatest virtue of NOA is to call attention to just how minimal an adequate philosophy of science can be" (1984, p. 101). It is still adequate, because the NOAer can offer explanations of interesting facts about science (such as the explanation of scientific mistakes that I developed in chapter 4, or perhaps even the explanation of novel predictive success in terms of truth).[1] It is minimal, because it comes without the metaphysical baggage.

However, there is another sense in which the natural historical attitude is not a skeptical attitude at all. In characterizing the NOA, Fine joins arms with the scientific realist in affirming our ability to have knowledge of unobservables. He writes:

> Do you or do you not believe in electrons and DNA (dinosaurs too, while we are at it)? I cannot answer for NOA since an attitude does not compel particular beliefs, but for myself the answer is an unhesitating yes. I take the

[1] See Blackburn 2005 (ch. 7) for a defense of the claim that someone who takes a minimalist approach similar to NOA can still go along with most of the realist explanation of the success of science.

question of belief to be whether to accept the entities or instead to question the science that backs them up. I have no reason to worry about the science in these and many other cases. . . . (Fine 1996, p. 184)

Likewise, someone who adopts the natural historical attitude affirms that we can and do have some knowledge of the unobservable past. All of my argumentation so far has rested on the assumption that we do know a good deal about the past. Notice that a radical skeptic who denies that we can know anything at all about the past would also have doubts about the two asymmetries: How can we be sure that we cannot manipulate the past? Maybe we can, but we have just not figured out how to do so yet. Our background theories tell us how historical processes destroy information, but how do we know that those background theories are correct? In stating at the outset that these two asymmetries are real, I was basically assuming that we do know something about the past.

The best way to appreciate the anti-skeptical aspect of the NOA as well as the natural historical attitude is to focus on what I will refer to as the principle of parity:

> *The Principle of Parity.* Scientific claims about unobservables that have been checked, double-checked, etc. and are well-supported by the evidence are just as credible as ordinary claims about observables that are well-supported by the evidence.

In other words, the observable/unobservable distinction does not mark any significant epistemological divide. (And this, of course, is exactly what scientific realists will say, too! Although Fine proclaims that "Realism is dead," he cannot possibly mean epistemological realism. This is also why he is sometimes accused of being a closet realist.) Since this parity principle is central to Fine's disagreement with van Fraassen, it is worth pausing to explore why someone might think that this principle is true.

A good argument for the parity principle, which Fine does not himself give, can be found in the writings of Michael Devitt (1991, pp. 149–150). Begin by considering a simple case of abduction, such as that involving footprints in the snow. When we see the tracks in the snow, we infer that a deer passed this way not long ago. The deer, of course, is no unobservable entity. In this case, we are drawing an inference from observed observables (i.e., the tracks) to the existence of an unobserved observable (i.e., the deer). Since we are still dealing entirely within the realm of the observable, presumably no one would find such an inference problematic. Suppose, however, we find fossil footprints, and infer that some creature – say a dinosaur – must have made them a very long time ago. This time, we are making an inference from observables (i.e., the tracks) to unobservables (i.e. to the existence of the dinosaur). Now Devitt argues that the only difference between the two inferences is that the second is an inference to the existence of an unobservable entity. Or to put it another way, the second inference carries us across the observable/unobservable divide, whereas the first does not. But since the inferences are otherwise exactly the same in both cases (same starting evidence, same inference rule), this fact alone cannot possibly provide us with good reason to reject the latter inference even while we endorse the former. Although he does not put it in quite these terms, Devitt's take-home message is that if we reject the parity principle, we will end up *treating similar inferences differently*, by accepting the conclusions of some but not others, and for no good reason. Although Devitt was one of my targets in the previous chapter – for he is wrong to suppose that the basic abductive argument lends any support to realism's mind-independence claims – his argument for the parity principle is more compelling than anything I have found in Fine's work.

In making this case for what I am calling the parity principle, I have focused on abductive inference, or inference to the best explanation, but this is not essential. The same argument can be stated in a general way, without any reference to abduction, by describing a case in which we first infer from some observed evidence, via inference rule R, that some unobserved but observable thing exists. Then we infer from very similar evidence, using the same rule R, that some unobservable thing exists. The parity principle just says that these similar inferences need to be treated alike, or that we should treat their conclusions as being equally credible.

Some of the things that Fine himself says about the realism debate suggest that he may wish to resist my interpretation of him as a realist along the epistemological axis of that debate, but an agnostic along at least two other axes (those having to do with metaphysics and the nature of truth). Indeed, he sometimes writes in a postmodern vein as if he wishes to dissolve the realism debate altogether, as if it were better not to raise general philosophical questions about science at all. However, this anti-philosophical strand of Fine's thought is somewhat in tension with his profession of trust in science. When he expresses his trust in what scientists say about unobservables, he is in effect taking sides with realists over and against selective skeptics such as van Fraassen.

7.2 EMPIRICAL ADEQUACY

In his classic, *The scientific image* (1980), Bas van Fraassen develops a philosophy of science that is founded upon a rejection of the parity principle. For this reason, van Fraassen is sometimes referred to as a *selective skeptic*: He thinks we can have knowledge of observables, but no scientific knowledge of unobservables. He agrees with the realist – and with Arthur Fine – that we should take scientific claims about unobservables literally, and that such claims are meaningful. This *semantic realism* is opposed to verificationism (the view that claims about unobservables are only meaningful insofar as they can be translated without remainder back into claims about observables) and instrumentalism (the view that statements about unobservables are meaningless tools for generating predictions).[2] He disagrees with the realist, however, concerning the aim of science as well as the attitude that we should have toward our best-confirmed scientific theories.

The most central concept in van Fraassen's view of science is that of *empirical adequacy*: A theory, he says, is empirically adequate just in case all of its observational consequences are true. Whereas the realist holds that the aim of science is to deliver theories that are true, even in what they say about the unobservable world, the constructive empiricist holds that the aim of science is merely to produce theories that are empirically adequate, or (roughly) theories that are true in everything they say about the observable world. Or again, realists will say that when we have a theory that is well-supported by the evidence, we ought to *believe* it, meaning that we ought to take it to be true. That goes not only for what the theory says about observables, but also for what it says about unobservables. The constructive empiricist says, by contrast, that instead of believing our theories, we ought only to *accept* them, where acceptance is a complex attitude that has both an epistemic dimension and a pragmatic dimension. To begin with the epistemic dimension, to accept a theory is simply to believe that it is empirically adequate. Pragmatically speaking, accepting a theory means showing a willingness to work with it, to explore

---

[2] Many authors, including Fine (1986), describe van Fraassen loosely as an instrumentalist. And van Fraassen's constructive empiricism does have a lot in common with old-fashioned instrumentalism: both stress predictive accuracy, and both views deny that we can have knowledge of unobservables, for instance. But there are differences, too. Van Fraassen (1980, p. 10) sides with realists in thinking that what theories say about unobservables should be "literally construed."

its consequences, to design experiments in light of it, and to subject it to further empirical tests.

The difference between scientific realism and constructive empiricism is easily brought out by focusing on an example of an empirically successful theory – say, quantum theory. Realists hold that we should believe what quantum physicists tell us about the microphysical world, even when what they say strikes us as counterintuitive or difficult to understand, and where belief means taking the theory to be true or at least approximately true. Constructive empiricists, on the other hand, hold that we are not entitled to believe anything that quantum theory says about the microphysical world at all; instead, we are to believe only that the theory is empirically adequate, which is to say that all the predictions it makes concerning the observable evidence are true. Since there might be more than one empirically adequate theory, we might ask whether it is possible to accept (in van Fraassen's sense) a number of different incompatible theories. If acceptance, as van Fraassen understands it, involved only the belief that a theory is empirically adequate, then we could indeed accept different incompatible theories at the same time. But acceptance has a practical component too: accepting a theory also means using it as the basis for one's further scientific work, relying on it to help frame questions for further research, letting it serve as a guide to experimental design, treating it as background knowledge, and so forth. This pragmatic dimension of acceptance at least makes it more difficult to imagine how scientists could accept two incompatible theories at the same time. Perhaps this is possible, though it would be a difficult feat to pull off: Try to imagine a scientist who lines up two different research projects which are guided by two different and incompatible background theories, both of which the scientist accepts. When working on the first project, the scientist treats *T1* as a fixed body of background assumptions; when working on the second project, the scientist treats *T2* as fixed.

The main thing which separates van Fraassen from the scientific realist is his selective skepticism about scientific claims concerning unobservables. This selective skepticism also separates him from the NOAer, since it conflicts with the principle of parity, discussed above. Were I to offer a more thorough discussion and evaluation of van Fraassen's constructive empiricism, it would be necessary, first, to consider his main arguments for selective skepticism, as well as his famous critique of abductive reasoning, and secondly, to examine some of the main objections that other philosophers of science have leveled against his view. But this would take us somewhat far afield, and my main goal here is simply to use constructive empiricism as a foil to help us appreciate the non-skeptical aspect of the NOA and the natural historical attitude. To this end, I will forego any further discussion of the arguments and objections in order to zero in on a question that van Fraassen himself neglected: What would it mean to be a constructive empiricist with respect to the scientific study of the past?

7.3 CONSTRUCTIVE EMPIRICISM IMPLIES SKEPTICISM ABOUT THE PAST

At first blush, constructive empiricism does not seem like a terribly radical philosophical view. One can remain agnostic with respect to scientific claims about the unobservable microstructure of the universe, and at the same time happily join with everyone else in affirming the existence of tables, chairs, and other middle-sized dry goods. The heartening thing about selective skepticism is that it seems to leave virtually all of our ordinary external world beliefs intact. Constructive empiricism lets us keep all of our beliefs about observables, and demands only that we suspend judgment when it comes to the various claims that scientists make about unobservables. Just as the ancient Pyrrhonian skeptics enjoined us to "go by the appearances," so van Fraassen enjoins scientists to "go by the observables." And even that does not seem too bad, because they may still accept theories that make reference to unobservable entities and events, even though they are entitled to believe only what those theories say about observables. Van Fraassen's selective skepticism is a very straightforward kind of knowledge empiricism, or a version of the view that our knowledge is strictly limited to what human beings can experience.

When we turn our attention to historical examples, however, constructive empiricism begins to seem far more radical than van Fraassen originally made it out to be. Consider the following argument:

P1. All of our knowledge is limited to that which we can observe.
P2. We cannot observe things which no longer exist or events which occurred in the past.
C. Therefore, we cannot know anything about past things and events.

As far as I know, Philip Kitcher (1993) is the only philosopher who has taken van Fraassen to task on these issues. But it is hard to see how van

Fraassen can avoid this outcome. To begin with, he cannot give up P1 without giving up on constructive empiricism altogether. To deny P1 would be to affirm that we *can* sometimes have knowledge of unobservables, but that is just what scientific realists (and NOAers) would say. The selective skepticism embodied in P1 is what is supposed to distinguish constructive empiricism from epistemological realism. Since the argument is valid, the only other option is to give up P2. A discussion of the arguments for and against P2 would take us back over the territory covered in chapter 3. Kitcher argues forcefully, however, that given van Fraassen's own conception of how best to draw the distinction between observables and unobservables, P2 comes out true.

Indeed, van Fraassen does not say much about the observable/unobservable distinction that might help him avoid P2. He says that the predicate "observable" is vague, but that we can still give clear-cut examples of observable things and unobservable things. He also tries to naturalize the observable/unobservable distinction by leaving it up to empirical science to tell us the limits of our observational capacities. But none of this really helps, since the only way to avoid P2 is to make plausible the claims that we can observe living dinosaurs, or that we can observe the battle of Waterloo. The only thing that van Fraassen does offer here is the following principle:

> X is observable if there are circumstances which are such that, if X is present to us under those circumstances, then we observe it. (1980, p. 16)

Notice the odd use of the phrase, "is present to us." Van Fraassen could say that a living dinosaur is observable because there are circumstances such that, if the living dinosaur is present to us under those circumstances – for instance, if it is out on the green right now – then we observe it. But *are* there such circumstances? Kitcher rightly points out that how we answer this question depends on what we think about science fiction scenarios involving time travel, about the possibility of alternative evolutionary histories in which humans coexist with dinosaurs, and other weird scenarios.

> Discussing the merits of far-fetched scenarios for bringing us into contact with extinct organisms strikes me as faintly ludicrous, a venture born out of a misguided test for judging our epistemic access to various aspects of nature. Paleontologists think they know a great deal about the past history of life. Why should the decision about whether their claims are true or merely empirically adequate turn on recondite scenarios which, even if possible, are never likely to be actualized? (Kitcher 1993, p. 153)

Must we give up our beliefs about dinosaurs, as constructive empiricism seems to require? That depends on whether dinosaurs are observable. But van Fraassen's own standard for observability makes this issue depend, in turn, on what we decide to say about certain "recondite scenarios." And this suggests, in turn, that we have pursued the inquiry down the wrong alley.

Past entities and events were not always unobservable, but they are now, and that means that the constructive empiricist must be prepared to endorse an extreme form of skepticism about the past. Had van Fraassen (1980) discussed historical science at all, or even raised the question what it might mean to be a constructive empiricist with respect to historical science, this consequence of his view would have become apparent to anyone. With the exception of Kitcher, van Fraassen's critics have let this problem slide. This is another way in which the scientific realism debate has been skewed by the failure to consider examples drawn from paleobiology and geology. Philosophical views which seem plausible in the context of experimental science begin to seem far less plausible when applied to historical science.

It should not be too surprising to learn that van Fraassen's constructive empiricism implies radical skepticism about the past. As we saw back in chapter 2, the way to defend radical skepticism about the past is to use a global underdetermination argument: For any claim P that scientists make about the past, God could have set things up to look as if P were true when he created the world a mere five minutes ago. We can use this algorithm to generate empirically equivalent rivals to any claim or set of claims about the past. Van Fraassen's case for constructive empiricism also depends crucially on using global underdetermination arguments in the context pf physics (1980, chapter 3). But why should philosophers of science be permitted to get away with using global underdetermination arguments in one context, while simply ignoring the unpleasant consequences that such arguments would have in another context?

7.4 A SENSE IN WHICH THE NATURAL HISTORICAL ATTITUDE IS "NATURAL"

So far, I have only argued that van Fraassen's constructive empiricism, when applied to historical science, leads to an extreme form of skepticism. This skepticism would threaten to sweep away not only our beliefs about prehistory, but also our beliefs about human history, and perhaps even our beliefs about what we had for breakfast yesterday. It is more than I can do here to give a full treatment of the problem of skepticism about the past. In chapter 2, I adopted, though only provisionally, the popular policy of dismissing radical skeptical hypotheses, such as the hypothesis that God created the world five minutes ago, on the grounds that they are not "genuine rivals" of our scientific theories. And I will stick by that policy. Nevertheless, I think it can be useful to imagine what it would be like for a constructive empiricist to admit that his selective skepticism about unobservables applies to prehistory, not to mention history. This will help to bring out one more important difference between constructive empiricism and the natural historical attitude.

Although the constructive empiricist would have to get rid of all of his beliefs about dinosaurs, he could replace them with weaker epistemic attitudes, or with beliefs about statements about dinosaurs. So for example, instead of believing that the (non-avian) dinosaurs went extinct approximately 65 million years ago, he might believe that the statement that they did is part of an empirically adequate overall theory about prehistoric life. Or he could substitute some other notion for empirical adequacy. For example, he could say that the statement that the dinosaurs went extinct 65 million years ago is part of the most coherent explanatory story that makes sense of all the available observational evidence, or something like that. Then he could combine this weaker epistemic notion with a pragmatic notion, just as van Fraassen does, by saying that scientists who accept (not believe!) the claim that the dinosaurs went extinct 65 million years ago will rely on that claim when confronting future research problems. So science can go on, but without any claim to having discovered any truths about the past at all.

One of the oldest complaints about extreme skepticism is that it is not psychologically possible, meaning that it is just not possible for creatures like us to live out our lives as radical skeptics. This was a favorite theme of David Hume, who observed that although radical skeptical arguments (such as global underdetermination arguments) seem unbeatable when we are considering them in the course of a philosophical discussion, it is impossible to remain convinced of their conclusions when the practical concerns of ordinary life take over. Is it even psychologically possible for someone to disbelieve in dinosaurs, or in trilobites? But perhaps van Fraassen can counter this classical worry about extreme skepticism by arguing that his notion of acceptance gives us just about everything we might want from belief, but without the epistemic overreaching. Maybe it would be impossible for anyone simply to jettison all their previously held beliefs about prehistory, but it might not be such a problem to replace one's former belief states with new acceptance states, or states that involve some weaker attitudes.

This reply to the worry about the psychological possibility of sustaining radical skeptical doubt only raises a new problem, however. Is it psychologically possible for anyone to accept a whole range of claims about the past without letting that acceptance slide into belief? The constructive empiricist would have to remain vigilant at all times to prevent this slide from occurring. The trouble is that when we look at the practical side of acceptance, the distinction between belief and acceptance begins to get a little blurry. Van Fraassen writes that "even for those of us who are not working scientists, the acceptance involves a commitment to confront any future problems by means of the conceptual resources of this theory" (1980, p. 12). So if we accept Darwin's theory, we rely on it when we confront practical problems, such as the appearance of antibiotic resistance in bacteria. Someone who accepts Darwin's theory will talk about the problem of antibiotic resistance using Darwinian language, will see the problem as one that involves an evolutionary process, and will offer policy proposals that are informed by Darwin's theory. Now how is this "commitment to confront any future problems by means of the resources" of Darwin's theory any different, at the end of the day, from taking Darwin's theory to be true? The constructive empiricist/Darwinist may constantly remind herself that she doesn't really believe any of Darwin's claims about the past; she only believes that they are empirically adequate. But since the practical dimension of acceptance is utterly indistinguishable from the practical dimension of belief, one wonders whether it is even psychologically possible for anyone to avoid the slippage from acceptance to belief.

Paul Horwich (1991) and Simon Blackburn (2005) make much the same point. However, instead of worrying about the psychological possibility of preventing acceptance from slipping over into belief, they worry more about the intelligibility of the acceptance/belief distinction to begin with. Blackburn begins by pointing out that the pragmatic dimension of acceptance, in van Fraassen's sense, involves immersion in a theory:

> Immersion, then, is belief in empirical adequacy plus what we can call being "functionally organized" in terms of a theory. This means being at home in its inference patterns and models. We are really talking of someone who has, perhaps tacitly, internalized the conceptual organization a theory involves.

> We could better call this *living* a theory, or . . . being *animated* by a theory.
> It is all of this that is still to be contrasted with belief. But is there a real
> distinction? (Blackburn 2005, p. 187, emphasis in the original)

Blackburn's point is a pragmatist one: Although one can draw a clear
enough theoretical distinction between believing that some theory is true
and believing only that it is empirically adequate, this distinction cannot
be sustained once the philosopher leaves the shade, not only for psychological reasons (that is the Humean point), but because it is a distinction
between types of psychological states that makes no practical or psychological difference. The states of acceptance and belief would play exactly
the same role in anyone's life.

One can take home another lesson from this discussion: Both constructive empiricism and the natural historical attitude have skeptical streaks.
Those who adopt the natural historical attitude will remain deeply skeptical of metaphysical claims about the past, and so will see the debate
between metaphysical realists and social constructivists in much the same
way that an agnostic might view a debate between theists and atheists –
that is, as a debate between two groups of dogmatists, both of whom make
claims that outrun the available evidence. The constructive empiricist, on
the other hand, entertains sweeping doubts about everything that scientists (or anyone) might say about unobservables. We are now in a position
to see that one interesting difference between the two views concerns the
naturalness of the doubts involved. Not only is it psychologically possible to suspend judgment on the metaphysical issues that scientific realists
and social constructivsts sometimes get worked up about, but doing so is
(I hope you will find) both easy and natural. The doubts that may arise
as a result of our contemplation of the asymmetry of manipulability and
the role asymmetry of background theories are also natural, since they
derive either from common sense, or else from background theories that
are developed in the course of scientific inquiry. In a word, the doubts
associated with the natural historical attitude come naturally, whereas the
constructive empiricist's blanket skepticism about unobservables represents a far less natural kind of doubt, and one that is imposed on science
from the outside.

7.5 TRUTH AND REFERENCE

Another aspect of Fine's NOA that has puzzled some commentators is
his view about truth and reference. Fine is a *semantic realist* – just like van

Fraassen, and just like scientific realists – since he holds that statements about unobservables are literally meaningful, even when they cannot be translated back into statements about observables alone. That much is uncontroversial. However, when Fine begins to talk about truth and reference, as in the following passage, he quickly begins to sound like a realist:

> When NOA counsels us to accept the results of science as true, I take it that we are to treat truth in the usual referential way, so that a sentence (or a statement) is true just in case the entities referred to stand in the referred-to relations. Thus, NOA sanctions ordinary referential semantics, and commits us, via truth, to the existence of the individuals, properties, relations, processes, and so forth referred to by the scientific statements that we accept as true. Our belief in their existence will be just as strong (or weak) as our belief in the truth of the bit of science involved, and degrees of belief here, presumably, will be tutored by ordinary relations of confirmation and evidential support. (Fine 1984, p. 98)

So the NOAer can still talk truth with the realist, can still say that the theoretical terms of our best supported theories genuinely refer, and can still believe in the existence of electrons, fractional electric charges, and so on. All of this sounds a lot like realism, especially when it is compared with constructive empiricism. The constructive empiricist, by contrast, will not talk truth with the realist (she will only talk of empirical adequacy), will not assert that central theoretical terms genuinely refer, and will not believe in the existence of electrons, fractional electric charges, and so forth.

So far, I have characterized the NOA (as well as the natural historical attitude) as an attitude of agnosticism towards metaphysics. But this leaves open the question of what the NOAer should say about the nature of truth.[3] Moreover, commentators have had considerable difficulty figuring out what Fine himself thinks about truth (see, e.g., Psillos 1999, chapter 10). I shall proceed by working backwards: If the very soul of the NOA is, as I have been arguing, neutrality along the metaphysical dimension of the scientific realism debate, then we should proceed by asking which views of truth might be compatible with that metaphysical neutrality. After all, the NOA is the natural *ontological* attitude, not the natural attitude about truth.

---

[3] For helpful introductions to the recent and current debate about truth, see Kirkham (1992) and Lynch (2004).

To begin with, it may help to consider the realist, or correspondence conception of the nature of truth.[4] According to that view, truth consists in accurate representation. That is, what makes a statement true is its agreement with reality, or the way the world is. All of this is rather vague, and proponents of the correspondence theory have had a difficult time trying to make it more precise. For present purposes, though, the important thing to see is that the correspondence theory of truth is compatible with metaphysical anti-realism, or with social constructivism, or with the view that things and events in the world metaphysically depend on us, our theories, and/or our conceptual schemes. Saying that truth consists in agreement with reality leaves it open whether the reality in question is mind-dependent or –independent.[5] So one possible view would consist in (a) epistemic realism; (b) realism about the nature of truth, or belief in the correspondence theory of truth; and (c) agnosticism about the mind-dependence or –independence of unobservables. This is probably not the view that Fine holds, but it is a view that he could hold, if he wanted to.

A second possible view, which I think is the one that Fine does hold, is agnosticism with respect to the nature of truth. An agnostic about the nature of truth would simply suspend judgment on the question of what it is, precisely, that makes true statements true and false statements false. Realists hold that agreement or disagreement with the world makes our statements true or false. Epistemic theorists might say, by contrast, that facts about what we would be justified in believing once all the evidence is in are what make our statements true or false. Coherence theorists would say that statements are true or false in virtue of their relations to other statements. But it is certainly open to someone who regards this philosophical debate as unsettled to remain agnostic about the nature of truth, and even to remain agnostic on the question whether truth has a nature to begin with.

Someone who remains agnostic about the nature of truth can still say a lot of things about truth that sound like what the realist would say. Most importantly, the agnostic can and should still endorse the T-schema:

"*P*" is true if and only if *P*.

---

[4] Some philosophers have suggested that commitment to the correspondence theory of truth is essential for being a scientific realist. Thus, Putnam writes that "whatever else realists say, they typically say that they believe in a 'correspondence theory of truth'" (1984, p. 140).

[5] I think that many contributors to the scientific realism debate fail to appreciate the compatibility of a realist/correspondence theory of truth with anti-realist metaphysics. I am greatly indebted to Michael Lynch for helping me to see this clearly.

This is what Fine has in mind in the passage quoted above, when he says that "NOA sanctions ordinary referential semantics, and commits us, via truth, to the existence of the individuals, properties, relations, processes, and so forth referred to by the scientific statements that we accept as true."[6] At the very beginning of this book I gave the following example of a statement about unobservables:

> When x-ray photons pass through a liquid sample that is thin compared to its x-ray absorption depth, less than 1% of the photons are scattered. (Anfinrud and Schotte 2005, p. 1192)

Anyone who adopts what Fine calls the "core position" will agree that this statement and a great many like it are true. From that, together with the T-schema, we can infer that there are x-ray photons, that liquid samples have x-ray absorption depths, and so on. This may sound realist, but it is really just a boring application of the T-schema to the core position. One can apply the T-schema in this fashion without taking on any further commitments about the nature of truth. One can even suspend judgment about whether truth has any nature for us to discover, above and beyond what is given by certain platitudes, such as the T-schema.

In order to mount a respectable defense of this agnosticism about the nature of truth, it would be necessary to go into much greater detail concerning the current debate about truth, and that would be a diversion from the main inquiry of this book. Let me gesture, though, in closing, at how an argument for agnosticism about the nature of truth might go, at least in the context of historical science: Our only empirical evidence about the past consists in observable historical records. Just as these records give us no evidence at all concerning the mind-dependence or – independence of past things and events, neither do they give us any evidence that would help settle a philosophical dispute about what makes our statements about the past true or false. Philosophical theories about the nature of

---

[6] The following quotation from Fine's classic essay has caused a lot of trouble:

> Rather, NOA recognizes in "truth" a concept already in use and agrees to abide by the standard rules of usage. These rules involve a Davidson–Tarskian, referential semantics . . . (1984, p. 101).

One problem with this is that many people may have some sort of pre-philosophical commitment to truth-as-correspondence, so that the concept already in use is a correspondence concept. My own suggestion for interpreting this passage is that Fine should have said "predicate in use" rather than "concept in use." The "standard rules," such as the T-schema, are rules governing the usage of the truth *predicate*.

truth for statements about the past are underdetermined by the historical evidence.

7.6 ANOTHER WAY IN WHICH THE REALISM DEBATE HAS BEEN SKEWED

Most philosophers of science today see van Fraassen's constructive empiricsm and Fine's NOA as the two leading philosophically respectable alternatives to scientific realism. Both Fine and van Fraassen were inordinately concerned with physics, and especially with quantum physics, and up to now no one (with the exception of Kitcher 1993) has paused to consider how either of these approaches to philosophy of science might apply to historical science. In this chapter, I have argued that Fine's view comes off looking much better than van Fraassen's. The NOA generalizes fairly easily to historical science, giving rise to what I have been calling the *natural historical attitude*. We are now in a position to give a complete characterization of that attitude:

i. Epistemic realism and commitment to the "core position": it is possible for us to have knowledge of the past. Indeed, we do have quite a bit of knowledge of prehistory.
ii. Agnosticism about the metaphysics of the past: maybe we have made prehistory, and maybe we haven't – who knows?
iii. Agnosticism about the nature of truth for statements about the past: We can say a lot just by applying the T-schema to the core position, but we can and should just suspend judgment about what makes our statements about the past true or false.
iv. A healthy respect for the role asymmetry of background theories and the asymmetry of manipulability.

This view is one that both scientists and philosophers of science can be for, and can make their own.

In stark contrast to Fine's NOA, van Fraassen's constructive empiricism does not generalize to historical science at all. Adopting the constructive empiricist view of historical science would mean becoming a radical skeptic about the past. But since we arguably cannot get rid of all of our beliefs about the unobservable past, constructive empiricism does not even enter the running as a serious philosophy of historical science. Earlier (in chapter 3) I argued that the realism debate has been skewed by the failure to consider examples from historical science, because the traditional abductive arguments for realism have less force in historical than in experimental contexts. Now we can see that the debate has been skewed in another way as well: one of the leading non-realist philosophies of science works equally well in both historical and experimental contexts, but the other does not.

There is one other aspect of Fine's approach to the philosophy of science which fits especially well with the main argument of this book. In a contextualist spirit, Fine strenuously opposes attempts to offer global philosophical interpretations of the scientific enterprise. For example, he repeatedly objects to the way in which both realists and constructive empiricists talk about *the* aim of science (whether truth or empirical adequacy), on the grounds that science is not the sort of thing that could possibly be organized around a single aim. Talk about the aim of science also presupposes that science has some sort of essence or nature that philosophers can discover – another view that Fine opposes. He writes that "NOA tries to shift the philosophical focus away from the global and toward the local, away from the general (or universal) and toward the particular" (1996, p. 179). He wants philosophers to "scale things down to contexts in which philosophical inquiry can thrive." Unfortunately, however, he also tends to be a bit vague about what such a localized philosophy of science might look like.

Yet if Fine really wanted to drive home this contextualist argument against global interpretations of science, and in favor of localized philosophical inquiry, he could not do better than to draw attention to the asymmetry of manipulability and the role asymmetry of background theories. The problem with the realism debate, as it has been carried out at the global level, is that interesting, epistemically relevant differences – in this case, differences between the past and the microphysical – have gone unnoticed. One good argument for taking philosophy of science local, as Fine recommends, is that the debate among the partisans of various global interpretations of science has been skewed in the ways I have suggested.

8

Snowball Earth in the balance

Sciences such as geology and paleobiology are collocative, in the sense that researchers in these fields typically try to gather as much evidence of as many different kinds as possible, and they try to devise coherent explanations that make sense of all the disparate evidential traces. Notions like consilience, explanatory coherence, and explanatory unification figure prominently in collocative science. But these notions also raise difficult philosophical questions: What is consilience, exactly? By what right may scientists take consilience as a guide to the truth about the past? What should someone who adopts the natural historical attitude say about the role of consilience in historical science? And what consequences might the asymmetry of manipulability and the role asymmetry of background theories have for the interpretation of appeals to consilience in historical science? No discussion of historical science would be complete without some treatment of the role of consilience.[1] In this chapter, accordingly, I begin with a case study from earth science and proceed from there to discuss the philosophical issues. I argue that while consilience does carry some evidential weight, the asymmetry of manipulability and the role asymmetry of background theories imply that the problem of equiconsilient models will be especially common in historical science. This continues the theme from earlier chapters, which is that the asymmetries place historical researchers at an epistemic disadvantage.

---

[1] Although I am focusing on a case study from geology, other authors – most notably Alison Wylie (1995; 2002, ch. 16) – have also discussed the importance of appeals to consilience in archaeology.

8.1 THE APPEAL TO CONSILIENCE IN THE SNOWBALL
EARTH DEBATE

The snowball Earth debate, which we have already encountered a couple of times in earlier chapters, can serve as an example of collocative science in action, and it is a good case study to work with for several reasons. First, one of the big problems with the snowball Earth scenario has had to do with the relationship between the alleged deep-freeze episodes during the Neoproterozoic and the evolutionary history of life on earth. The Cambrian explosion occurred around 550 million years ago, just after the snowball Earth episodes are thought to have occurred. Did global glaciation somehow set the stage for the Cambrian explosion? And how could any life at all have survived the snowball Earth episode? The persistence of life through a series of extreme "freeze and fry" episodes has always posed a challenge to proponents of snowball Earth. For now, the important thing to see is that the snowball Earth debate has both geological and paleobiological dimensions. In some cases, scientists have offered biological explanations of geological phenomena. For example, Kirschvink and colleagues (2000) explain the existence of the Kalahari manganese field in southern Africa in terms of "global glaciation followed by a cyanobacterial bloom."

Another reason why the snowball Earth debate makes for a useful case study is that the history of the snowball Earth scenario eerily parallels the history of other geological theories, such as continental drift. From the 1960s on, few in the geological community took the idea of global glaciation seriously, in part because no one understood how the mechanism of deglaciation could work. Then Kirschvink (1992) addressed this problem by pointing out that volcanism would continue even if the entire planet were covered in ice. Volcanoes would have spewed carbon dioxide into the atmosphere, and the ice pack would have disrupted the carbon cycle, so that the carbon dioxide would have had nowhere else to go. Over millions of years, carbon dioxide would have accumulated in the atmosphere up to the point where a powerful greenhouse effect would have rapidly melted the ice. Solving this problem concerning the mechanism of deglaciation paved the way for a more serious consideration of the snowball Earth scenario during the 1990s. Similarly, when Alfred Wegener first suggested the theory of continental drift, many geologists dismissed it out of hand because no one could imagine how the continents could slice through the solid ocean floor. Only much later with the development of the theory of plate tectonics did scientists begin to take continental drift seriously, because the theory of plate tectonics provided an underlying mechanism for the movement of the continents, much as Kirschvink provided a mechanism for deglaciation after a snowball Earth episode.

Building on the work of Kirschvink (1992), Hoffman et al. (1998) published a paper in the journal *Science* in which they showed how the snowball Earth theory could unify a number of disparate and otherwise puzzling historical traces. These traces include: (1) the evidence of low-latitude glaciation during the neoproterozoic; (2) the fact that in many places where these glacial deposits are found, they occur right underneath a thick layer of carbonate rock, which is known to form only on seafloors at warm temperatures; (3) the fact that neoproterozoic glacial deposits contain rich deposits of iron, which could only have formed under unusual conditions, because iron can only dissolve in water when there is no oxygen present; and (4) the unusual ratios of carbon-12 to carbon-13 isotopes in neoproterozoic rocks, which suggest that photosynthetic activity in the earth's oceans shut down for a period and then rebounded. In the nineteenth century, William Whewell characterized consilience as a "jumping together" of inductions, and that, roughly, is what happened here: Hoffman and Schrag showed how to explain these four distinct phenomena as consequences of a series of global ice ages. In the years since 1998, more and more scientists have begun working on this problem, and some parts of the theory advanced by Hoffman and Schrag have been challenged, though by now nobody seriously doubts that there were glaciers of some sort in the tropics during the neoproterozoic. This attention has been due in large part to the effective appeal to consilience that Hoffman and Schrag made in their 1998 paper. In a popular article that came out soon afterwards, they write of snowball Earth that "the strength of the hypothesis is that it simultaneously explains all of these salient features, none of which had satisfactory independent explanations" (2000, pp. 7–8).

Kirschvink (1992) is actually the one who first showed how the snowball Earth hypothesis might explain the iron-rich deposits. The neoproterozoic glacial deposits contain large quantities of iron-rich rock, and it is hard to understand how so many iron-rich deposits could have been formed at one time. Knowing that iron is not ordinarily soluble in the presence of oxygen, Kirschvink proposed the following explanatory story: Over the course of millions of years, the world's oceans, which were covered by a thick layer of ice, became deoxygenated. Meanwhile, iron released by steam vents in the ocean floor gradually accumulated in the seawater. When the ice finally melted, a huge quantity of iron came into contact with the oxygen in the atmosphere and precipitated out along with all the glacial debris.

What about the strange carbon isotope ratios? The main sources of carbon in seawater are undersea volcanoes and steam vents, and carbon-12 and carbon-13 isotopes are released from those sources in a fixed proportion (about 1 percent carbon-13). However, the photosynthetic activity of marine organisms, especially algae, alters that proportion, because photosynthesis uses up more of one isotope than the other. Rocks in Namibia and elsewhere tell the following story: During the Neoproterozoic, the carbon isotope ratio fell nearly to the baseline – that is, to the fixed ratio that one would expect to see if there were no photosynthetic activity in the earth's oceans. This ratio persists through the glacial deposits and into the cap carbonate rocks above, where the amount of carbon-13 begins to increase again from the baseline of 1 percent or so. Hoffman and Schrag argue that this can be explained by supposing that the earth's oceans were trapped under a layer of ice. Photosynthetic activity would have ground to a halt during a snowball Earth episode lasting perhaps millions of years and then ramped up again when the sea ice melted.

Hoffman and Schrag then proceed to offer an explanation of the cap carbonates. They argue that the thick layer of carbonate rock is exactly what one would expect to see if a long snowball Earth episode were suddenly brought to an end by runaway global warming due to huge buildup of carbon dioxide in the atmosphere. They tell a story about radical climate change: First, the ice melts. As the area of the earth's surface covered by ice decreases, the albedo effect decreases, and the planet warms up even more, with average global temperatures possibly reaching 50 degrees Celsius. Huge amounts of water evaporate. Rainstorms and hurricanes wash the carbon dioxide out of the atmosphere in the form of carbonic acid. Acid rain erodes newly exposed rocks, and the products of this erosion eventually end up on the seafloor in the form of carbonate sediments. Large quantities of carbonate sediment thus end up sitting right on top of debris that was dropped into the ocean by melting sea ice. In this way, the snowball Earth hypothesis weaves together the glacial debris, the iron-rich deposits, the cap carbonates, and the carbon isotope ratios, and perhaps even the subsequent Cambrian explosion into a single coherent geological story.

The snowball Earth hypothesis thus explains a diverse set of geological traces. What should we make of Hoffman and Schrag's appeal to consilience, in the light of everything that has been said about the asymmetries of manipulability and the role asymmetry of background theories? What should someone who adopts the natural historical attitude think about such an appeal?

As we saw in connection with the discussion of underdetermination in chapter 2, philosophers of science typically distinguish between empirical and non-empirical theoretical virtue. Empirical virtue is simply a matter of having true observational consequences. We can imagine two empirically equivalent theories which for that very reason have equal doses of empirical virtue, but which nevertheless differ when it comes to other virtues, such as simplicity, explanatory power, coherence with other theories, and so on. It is customary to refer to these other virtues as "non-empirical" or "supra-empirical," since they seem to involve features of theories that are not just a matter of those theories' empirical consequences. Traditionally, consilience has been classified as a non-empirical theoretical virtue, or a criterion of theory choice that scientists appeal to when theory testing becomes especially difficult – as is so often the case in historical science. Here it is worth remarking that almost no one in the geological community responded to Hoffman et al.'s important 1998 paper by saying, "Now the snowball Earth hypothesis has passed its first major empirical test." Instead, Hoffman and Schrag's paper was interpreted as an argument for taking the snowball Earth hypothesis very seriously, and for seeking new ways to test it (See, e.g. Kerr 2000).

Suppose that some other geologists – professional rivals of Hoffman and Schrag – were to introduce four independent explanations of the puzzling phenomena mentioned earlier: the glacial deposits, cap carbonates, iron-rich deposits, and carbon isotope ratios. Instead of offering a unified explanation of these data, the rivals offer four distinct and unconnected explanations (call them E1, E2, E3, and E4). But if we so choose, we can still conjoin these four unconnected explanations to form a "theory" of sorts: E1 and E2 and E3 and E4. With respect to empirical virtue, we now have a tie, although it is a tie that could be broken when additional evidence comes in. The data gathered so far are implied by both theories, and so cannot be used to discriminate between them. However, the snowball Earth hypothesis wins on account of the fact that it is more consilient. It offers one unifying explanation, whereas the rival "theory" offers four distinct explanations that are merely conjoined together. This rival "theory" offers what some might call a *spurious unification*. We can say, if we want to, that the conjunction "E1 and E2 and E3 and E4" unifies the evidence, but this is just a trick, because the conjuncts (let us suppose) are not relevant to one another at all. This example also shows that the degree of consilience of a theory is not simply a matter of the empirical consequences of that theory.

What is consilience, exactly? For present purposes, I will work from the assumption that when philosophers use terms such as "consilience," "explanatory power," and "explanatory coherence," they aim to refer to the same theoretical virtue. There may be subtle distinctions between them (as there may be a subtle distinction between courage and bravery), but for present purposes those subtle distinctions will not matter. Still, given that the appeal to consilience has played such a central role in the snowball Earth debate, it would help to be able to say with greater precision what consilience is.

8.2 GOING BEYOND "SEAT-OF-THE-PANTS FEEL"

Hilary Putnam once wrote that since we do not have any algorithm that will tell us how coherent a theory or belief system is, we must rely on "seat-of-the-pants feel" when making judgments about coherence (1981, p. 133). Unfortunately, there are some serious problems with relying on seat-of-the-pants feel. What happens when two scientists, or two research groups, both relying on seat-of-the-pants feel, arrive at conflicting judgments about which of two models is more consilient? How should we adjudicate such disagreements? If we are unable to say with any precision what consilience (or coherence) amounts to, then how could we ever justify treating consilience as a guide to the truth? What could possibly justify using seat-of-the-pants feel as a guide to truth? In a recent article that challenges philosophers to explain just what they mean by "coherence," Elijah Millgram writes that "A coherence concept that the troops cannot use in the field is also one whose adoption it will be next to impossible to justify" (2000, p. 83). What we need is a consilience concept that the troops can use in the field. Without such a concept, it will be difficult to make much sense of Hoffman and Schrag's argument for the snowball Earth hypothesis.

Other philosophers of historical science have failed to meet Millgram's challenge. For example, in his recent book on the philosophy of archaeology, Peter Kosso defends a coherence theory of knowledge of the past. In keeping with tradition, he defines coherence in terms of mutual support relations:

> The justification must be in the mutual relation, the coherence of the network, rather than a basic terminus of support. One of the ingredients of coherence, and the easiest to evaluate, is consistency. A description of events and things in the past must be free of contradiction. But more than that, a system of claims coheres to the degree that there is explanatory relevance among various claims. (Kosso 2001, p. 75)

This is not yet a coherence concept that the troops can use in the field. The phrases "a system of mutually supporting claims," and "a system of claims bound together by explanatory relevance relations," are, to be sure, roughly synonymous with "a coherent set of claims." But if we ask how we are supposed to assess the degree to which "there is explanatory relevance among various claims," the answer will still be seat-of-the-pants feel.

Kosso adds, however, that coherence, as defined above, is necessary but not sufficient for epistemic justification. In order to complete the picture, he further develops his coherence theory of archaeological knowledge in two ways. First, he defends a weighted coherence theory, according to which "some claims in the network are epistemically weightier than others and are less likely to be challenged or abandoned" (Kosso 2001, p. 92). This is plausible: Critics of the snowball Earth model have treated some components of that model as having more weight than others. For example, few have challenged the bare claim that there were glaciers in the tropics during the neoproterozoic, though other details of the snowball Earth model have met with sharp criticism.

Next, Kosso requires that "in the relation of testing theory against evidence, the theory that gives meaning and credibility to the evidence must be epistemically independent of the theory being tested" (2001, p. 92). At some points it seems like his conception of epistemic independence is very close to the scientific realists' independence condition for predictive novelty (see chapter 5). However, it is not at all clear how epistemic independence in that sense will help us to understand appeals to coherence/consilience in cases where no novel predictions have been made – as for example in Hoffman and Schrag's (1998) argument for the snowball Earth hypothesis. We can safely assume that the independence condition on novelty is not the kind of independence that Kosso has in mind here.

Kosso's main motivation for emphasizing epistemic independence, and for developing a theory of justification based on "dynamic coherence with independence" is to avoid the traditional circularity objection to coherence theories of epistemic justification: According to such theories, a claim P receives what justification it has from its participation in a network of mutually supporting claims. But those other claims which support P also receive some of their justification from P. This looks like vicious circularity. Kosso defines epistemic independence with this objection in mind:

> One claim x is independent of another y in this epistemic sense just in case y does not entail any of the justification claims used to support x. Thus, if y does not contribute to the credibility of x, x can be used as independent evidence for y without incurring the problematic circularity of x supporting y while y supports x.[2] (Kosso 2001, p. 84)

The first thing to notice is that this is much weaker than the independence condition on predictive novelty. Take, for example, the claim that there is glacial debris in neoproterozoic strata. That claim is independent (in this weak sense) of the claim that the entire Earth froze over during the neoproterozoic. We have ample empirical justification for the claim that there is glacial debris in neoproterozoic strata without having to make any assumptions about whether the snowball Earth model is true. So the evidence that Hoffman and Schrag present for that model is epistemically independent of the model, in Kosso's sense. Pointing this out, however, does not help us get beyond seat-of-the-pants feel at all. In this case, the evidence is independent (again, in Kosso's weak sense) of the proposed theory. What we want to know is what it means to say that the snowball Earth theory offers a coherent/consilient explanation of that evidence. Although Kosso is right about many things – about the importance of coherence in archaeology, about the value of epistemic independence, and much else – it is not clear that he has yet answered Millgram's challenge. To put the point another way: We can agree that epistemic

---

[2] On a more technical note, I have doubts that this independence constraint is even compatible with a strict coherence theory of justification. Someone who is sympathetic to the circularity objection to coherence theories of justification could just argue that such theories automatically violate this independence constraint. I am inclined to think that Kosso's theory of "coherence with independence" is best interpreted as an example of what Susan Haack (1993) calls a foundherentist theory. Indeed, Kosso (2001, pp. 150–151) speaks favorably of Haack's view. And it is no coincidence that the circularity objection is one of Haack's main motivations for defending foundherentism rather than a strict coherence theory.

independence, in his sense is necessary for coherence, but what are the other ingredients?

There are at least two fundamentally different approaches to the problem of defining a term such as "consilience" or "coherence": An *exemplar approach* and an *analytical approach* (or alternatively: an extensional and an intensional approach). Virtually all philosophers of science who have tackled this problem have gone for the analytical approach. Fascinatingly, though, the late Stephen Jay Gould pursued the exemplar approach.[3]

The difference between these two approaches is related to the difference between two styles of definition: ostensive (or extensional) *vs.* intensional definition. One way of defining a term is to point to an example of something to which the term refers. To give a trivial example, one can define "dog" by pointing at Fido and saying, "That's a dog." But one can also define "dog" by giving the intension of the term – that is, by listing the attributes that people typically have in mind when they think of dogs: wagging tails, keen noses, loud barking, drool, and so on. This same point also applies to terms of art, such as "consilience." One way to define that term is to point to familiar examples of theories and models and say, "Now *that's* what I'm talking about when I say that a theory is consilient." We could say that Darwin's evolutionary theory, the theory of plate tectonics, and a few others serve as exemplars of consilience. Having picked out one or a few particular cases, we can then proceed to say that other historical theories and models are consilient insofar as they resemble those exemplars of good historical science. Those exemplars set the standard that scientists like Hoffman and Schrag are invoking when they play up the consilience of their own preferred models or theories.

Stephen Jay Gould (2002) is a good example of a theorist who appreciates the importance of appeals to consilience in historical science and who adopts what I am calling the exemplar approach. Gould portrays Charles Darwin as a methodological pioneer, and as the scientist who demonstrated just what it means in practice to treat consilience as an evidential standard. One could go on all day – and Gould does with great affection – about the various ways in which Darwin unified seemingly disparate phenomena. For example, Darwin showed that the process by which farmers and breeders, over the course of many generations, have produced new

---

[3] Another important source of inspiration for the exemplar approach is Thomas Kuhn (1996).

varieties of cattle, dogs, and pigeons, is essentially the same sort of historical process as that which has given rise to plant and animal species in nature. The same theory that explains interspecific differences (for example, between humans and chimpanzees, or between zebras and horses), can explain the existence of organs that appear to be fitted for some purpose. The theory can explain why, after years of applying a certain pesticide to their fields, farmers will begin to see insects that are immune to it. The theory can also explain interspecific similarities (for example, why are humans genetically so similar to chimpanzees?). What's more, Darwin unified the evidence from the fossil record with the visible patterns of similarity and difference in living organisms. According to Gould, Darwin's theory serves as a standard-setting example of a consilient historical theory. This, perhaps, is one sense in which historical scientists are all still working within the "Darwinian paradigm." Darwin's theory, however, is not the only standard-setting theory that one could point to. In geology, the theory of plate tectonics, though a relative newcomer, may play much the same role. The well-established theory of Pleistocene glaciation could also serve as a good exemplar of consilience in the context of the snowball Earth debate.

The point is simply that one can give a reasonably good extensional definition of "consilience" – and one that the troops can use in the field – just by pointing at such exemplars in the way that Gould does and saying, "There: *That's* how to make a convincing appeal to consilience. *That's* how great scientists unify the phenomena." Since all the members of the scientific community will have studied these exemplar theories as part of their professional training, such appeals will readily be understood by everyone. Moreover, the degree of consilience of competing models can be assessed simply by comparing them to the exemplars with which all members of the scientific community are intimately familiar, and which serve as a common standard for all. Such comparisons lack the arbitrariness that is implied by Putnam's phrase, "seat-of-the-pants feel."

Those who instead adopt the analytical approach prefer to begin by giving an analysis of the concept of consilience, or coherence.[4] Kosso himself goes in this direction when he defines coherence (albeit somewhat vaguely) in terms of the number and density of explanatory relevance relations in a network of claims. Without question, though, the philosopher who has done the most to develop an analysis of explanatory coherence is

---

[4] I would count Michael Friedman's (1974) and Philip Kitcher's (1989) work on explanatory unification as examples of this analytic approach, along with McGrew's (2003) recent attempt to reconstruct appeals to consilience in terms of Bayesian confirmation theory.

Paul Thagard. Since Thagard (1991, 1992) has also applied his coherence theory to the problem of conceptual change in geology, I shall briefly sketch his approach and explain why I do not find it – or any analytical approach, for that matter – to be very helpful.[5]

Thagard and Verbeurgt (1998) think of coherence in terms of the following task, which one could write a computer program to solve: Start with a set of elements – say, a set of claims, including some claims about the present evidence and some claims about past things and processes – and divide that set of elements into an *accepted set* and a *rejected set*. Next, suppose that the different ways of partitioning the initial set of claims into the accepted set and the rejected set are subject to various constraints. To begin with, there are *positive constraints*, which say that if one thing is included in the accepted set, then another thing should be included, too. These positive constraints can be thought of as representing relations of implication and/or positive explanatory relevance. Next, there are *negative constraints*, which say that if one thing is included in the accepted set, another thing should be excluded. These negative constraints can be used to represent relations of inconsistency and negative relevance. Finally, each constraint gets a numerical index, which reflects its weight, or its relative importance. (So far, this is very much in line with Kosso's weighted coherence theory.)

For any initial pool of claims, there will be vastly many different ways of subdividing them into an accepted and a rejected set. One could write a computer program that would look at each one of these possible partitions, one after the other. For each partition, the program could add up the weights of all the constraints that are satisfied by it. The resulting sum would serve as a numerical measure of the degree of coherence of that partition. Then, after looking at all of them, the computer could determine which of all the many possible partitions is the most coherent one, in the sense of satisfying the greatest number of weighted constraints. Thus, Thagard and Verbeurgt take some of the vague ideas that we find (for example) in Kosso's discussion of coherence in archaeology – consistency, explanatory relevance, weighted coherence, and so on – and operationalize them in a way that makes coherence something that can be computed. Rather than relying on seat-of-the-pants feel, we can just calculate.

It is unlikely that this computational approach will ever shed much light on historical science, however. Although it does give us a way of making

---

[5] Thagard (1991) applies the computational theory of coherence to the debate about the Cretaceous–Tertiary mass extinction. Thagard (1992) brings a similar approach to bear on the Darwinian revolution and the plate tectonics revolution.

Kosso's conception of coherence more precise, the precision comes at a high price. To begin with, Millgram (2000) argues that Thagard and Verbeurgt's approach runs into serious problems having to do with computational complexity. The computations required by this approach certainly could not be performed by the troops in the field, and they could get to be so complex that even the most powerful computers could not perform them. Relatedly, this model is just descriptively inaccurate: When Hoffman and Schrag argue that the snowball Earth hypothesis does a good job unifying the evidence, it is hard to believe that they are carrying out computations such as Thagard and Verbeurgt describe. It is also worth pointing out that this computational approach is subject to the old "garbage in, garbage out" problem: The output (a determination of the most coherent partition) will always be only as useful as the inputs (an initial set of claims, and a set of weighted positive and negative constraints). So even if scientists were able to perform the necessary computations – or if they had computers that could do it for them – then they would have to make tough decisions about what to use as inputs. And how would they make those decisions – for example, decisions about how to weight the various constraints? By seat-of-the-pants feel? One begins to get the sense that Thagard and Verbeurgt achieve the appearance of precision mainly by leaving all the difficult and messy aspects of the problem of coherence out of their model. Once we are given a set of inputs, we can easily understand how a program could go about determining the most coherent partition. The tough part is deciding what inputs to use – i.e. which constraints to impose, and how to weight them.

So far, at least, the analytical approach to the problem of defining consilience/coherence has not born fruit.[6] Indeed, I suspect that the analytical approach is subject to the following dilemma: Either we define "coherence" and "consilience" in a way that is too vague to answer Millgram's challenge, or else we give a more precise analysis of those terms while resigning ourselves to descriptive inaccuracy, if only because the scientists who appeal to these notions do not have any sophisticated philosophical analyses in mind. Kosso is stuck on the first horn of this dilemma; Thagard on the second. The best course of action is to avoid the dilemma altogether by sticking with the exemplar approach to defining consilience.

---

[6] Although I will not develop the argument in any detail here, I suspect that other versions of the analytical approach, such as Kitcher's (1989) formal account of the notion of an "explanatory store," will likewise turn out not to be very helpful to the troops in the field.

One interesting (and I think true) consequence of the exemplar approach is that consilience might mean different things to scientists working in experimental *vs.* historical science, because they might well point to different exemplars of that virtue. The very nature of consilience could vary from one context to the next. Philosophers of science should not expect to be able to give any global, context-neutral definitions of "consilience", but there is plenty of work to be done at the local level by looking at which exemplars stand in the background of particular appeals to consilience.

8.3 IS CONSILIENCE MERELY A PRAGMATIC VIRTUE?

Even assuming that it is possible to give a workable extensional definition of consilience by treating Darwin's theory as an exemplar, we must still ask how, if at all, consilience is related to truth. This is closely linked to the question we encountered back in chapter 2, whether it is ever reasonable to appeal to notions like simplicity and explanatory power in order to break evidential ties between empirically equivalent hypotheses. It is also related to the whole issue of inference to the best explanation – one of the most contentious issues in the realism debate – because consilience is one of the standards that one could use to determine which explanation is the best. Asserting that consilience is a reliable guide to truth amounts to endorsing inference to the best explanation as a reliable inference rule.

At one extreme, constructive empiricists will urge us to take a radically skeptical view of consilience. According to this view, consilience is a *pragmatic* rather than an *epistemic* virtue. The degree of consilience of, say, the snowball Earth hypothesis, gives us no reason at all for thinking that the hypothesis is true. At most, it gives us a good reason to accept that hypothesis, where acceptance involves belief that the hypothesis is empirically adequate together with a practical commitment to confront future problems in terms of the hypothesis. Or to put the point in a slightly different way, suppose we have an empirical tie between two incompatible hypotheses, in the sense that the two have made all the same predictions so far, but where one is more consilient than the other. The constructive empiricist would urge us to accept the more consilient hypothesis without taking it to be true, or even likelier to be true than not.[7] According to

---

[7] For a similar view, see Post (1987, pp. 70–72).

this view, consilience is just one feature that we would like our theories to have, and for purely practical reasons. Notice also how the constructive empiricist would interpret Hoffman and Schrag's appeal to the explanatory power of the snowball Earth hypothesis. The fact that the hypothesis unifies diverse phenomena is no reason for thinking it to be true, but it is a reason for (pragmatically) accepting the snowball Earth hypothesis, for working with it, subjecting it to further tests, and confronting new problems in terms of it. Thus, the constructive empiricist offers a way of understanding the importance of consilience without supposing that consilience (or non-empirical theoretical virtue more generally) carries any evidential weight at all.

Why be so skeptical? To start with, notice that the connection between consilience and truth (if there is such a connection at all) could come in varying strengths. For example, one could hold that a certain degree of consilience guarantees truth, or (what is a weaker view) that the most consilient of all the available theories is likelier to be true than not, or (a still weaker view) that the most consilient of all the available theories is likelier to be true than any of its rivals. The worry one might have, even about this weakest of the three views, is that such a view would only make sense if we could be sure that the world itself is somehow consilient. That is, it would only make sense to treat consilience as a guide to truth, in the context of historical science, if we knew already that the past was consilient. But how could anyone possibly know that? How does anyone know that the past is not messy and complicated, as opposed to consilient? Obviously, we cannot simply say, on pain of circular reasoning, that we know that the past is consilient because that is the most consilient hypothesis. So what is the evidence for the claim that the past is consilient?[8]

It is not clear, however, that these skeptical worries about consilience are ultimately sustainable. First, the distinction between accepting the most consilient hypothesis and actually believing it threatens to collapse, as I argued in chapter 7. Second, even if the distinction itself is a good one, it might not be psychologically possible for anyone to avoid slipping from acceptance into belief. Third, one could still *accept* the hypothesis that the past is consilient, in the sense of (a) taking that hypothesis to be empirically adequate, if not taking it to be true, and (b) resolving to confront future problems using the resources of this hypothesis. But

---

[8] For a helpful discussion of this problem, see Lipton (1991), who refers to it as "the Voltaire Objection" to inference to the best explanation.

anyone who takes the pragmatic dimension of acceptance at all seriously, and who sincerely accepts the hypothesis that the past is consilient, will subsequently treat consilience as a guide to truth. What else could it possibly mean to confront future problems in terms of the hypothesis that the past is consilient?[9] This shows, incidentally, that van Fraassen's constructive empiricism is unstable. His skepticism about unobservables seems to require treating non-empirical theoretical virtues as merely pragmatic, but it might be possible to bootstrap one's way up to the view that they carry evidential weight, using only what constructive empiricism gives us to work with – namely, the notion of acceptance.

Here is how the bootstrapping procedure would work in the Hoffman and Schrag case: Those scientists could begin by merely accepting the hypothesis that consilience is a reliable guide to the truth about the past. That is, they come to believe that this hypothesis is empirically adequate, and they resolve to confront future problems in terms of it. One of those future problems, it turns out, is the problem of how to explain the cap carbonates, iron-rich deposits, carbon-isotope ratios, and the other puzzling features of neoproterozoic rocks. But what could it mean to confront this problem in terms of the hypothesis that consilience is a reliable guide to truth? Any scientists who accept (in van Fraassen's sense) the hypothesis that consilience is a reliable indicator of truth will take the consilience of the snowball Earth hypothesis to carry some evidential weight. If scientists did not treat consilience as having evidential weight, then we should conclude that they do not really accept the hypothesis that consilience is a guide to truth in the first place.

Notice that this bootstrapping procedure does not involve a viciously circular argument – or really any argument at all. No one is arguing that we should believe that consilience is a reliable guide to truth about the past, on the grounds that this it iself a consilient hypothesis. One need only begin by accepting that hypothesis, in van Fraassen's sense of acceptance. Perhaps van Fraassen would try to block the procedure right at the beginning, by arguing that we are not entitled even to accept the hypothesis that consilience is a guide to truth. But the epistemic dimension of acceptance is so weak that this objection will not be very convincing. To accept a hypothesis (epistemically, speaking) is to say, "Everything this hypothesis says about the observable evidence is true." The hypothesis that consilience is a reliable guide to the truth about the past seems to meet this first condition of acceptability. The hypothesis that consilience

---

[9] This argument is inspired by Blackburn (2005).

is a reliable guide to truth also seems to meet the second, pragmatic condition of acceptability. One could argue that this hypothesis is especially useful because of its methodological implications.

8.4 REDUCING NON-EMPIRICAL VIRTUE TO EMPIRICAL VIRTUE

In his discussion of cladistic parsimony, Elliott Sober (1988) argues that appeals to non-empirical theoretical virtue (such as cladists' appeals to parsimony, or Hoffman and Schrag's appeal to consilience) should be interpreted as covert appeals to well-supported background theories. That is, cladists' talk about parsimony is just a convenient way of talking about certain background assumptions about the nature of evolutionary processes. Parsimony carries evidential weight only to the extent that those background process assumptions are well-supported by the empirical evidence. Notice that this is a reductionist strategy: Sober's idea is that the evidential force of non-empirical virtue can be rendered unproblematic by being reduced to the empirical virtue of certain background theories. Talk about the non-empirical virtue is, as he puts it, a "surrogate" for talk about those background theories. Perhaps this strategy could also help to make sense of the notion of consilience.

Throughout this book I have talked at length about how the background theories of historical science describe the processes by which information about the past is destroyed. But those theories also have a lot to say about how information is preserved in the fossil and geological records. It is not too implausible to suppose that historical researchers also rely on some highly general background theoretical assumptions about the ways in which information about the past is preserved. Perhaps something like the following is what lies behind Hoffman and Schrag's appeal to consilience:

> *The Disparate Trace Hypothesis.* When token events occur, they usually leave not just one but several different kinds of effects, or traces.[10]

---

[10] The disparate trace hypothesis is very close to one half of the asymmetry of overdetermination – viz., that earlier events are usually overdetermined by their effects. And my claim that the disparate trace hypothesis underwrites appeals to consilience in historical science resembles Cleland's (2002) claim that the asymmetry of overdetermination underwrites the methodological differences between historical and experimental science. But there is still a crucial difference between my view and hers: I don't think that the disparate trace hypothesis compensates much, if at all, for the asymmetry of manipulability and the role asymmetry of background theories.

Of course, it is important to be careful in assessing the epistemological consequences of this hypothesis. For if the arguments of this book are correct, then the disparate traces left by any given event may well be undetectable by us, and historical processes may destroy most of them in short order. Notice, though, that one can easily test the disparate trace principle in experience. If you hear about a car accident, for example, you might predict that it left disparate traces: skid marks on the pavement, a twisted guardrail, cuts and bruises, broken glass, higher insurance premiums, and so on. Such predictions are readily testable, and the disparate trace hypothesis seems well supported by ordinary experience. Notice that the disparate trace hypothesis does not say that anyone will actually find the traces in question, or that the traces will be easy to identify, or that they will survive for very long. It only says that token events usually leave traces of different kinds.

The disparate trace hypothesis could also underwrite some appeals to consilience. If the hypothesis is correct, that has implications for how scientists should proceed when they find a collection of individually puzzling traces, such as the glacial deposits, iron-rich deposits, cap carbonates, and carbon isotope ratios. If in general, token events usually leave traces of different kinds, then it is rational to inquire whether a collection of traces of different kinds may all be attributable to some earlier token event. If scientists do succeed in unifying the traces by connecting them with a single event, then they can be somewhat confident that they are on the right track, because they would have told what, according to the disparate trace principle, is the right sort of explanatory narrative. Consilience does have some evidential weight, but appeals to consilience are best interpreted as shorthand for appeals to empirically testable background assumptions, of which the disparate trace hypothesis is just one example. If consilience gives us any reason to believe in snowball Earth, that is only because the disparate trace hypothesis (or some similar background assumption) has a great deal of empirical support. Thus, what van Fraassen arguably fails to appreciate is the way in which apparently non-empirical virtue can actually be reduced to the empirical virtue of background theories along the lines suggested by Sober.

I do not want to suggest that all appeals to consilience should be understood as tacit appeals to the disparate trace hypothesis. Instead, all I mean to suggest is that if we want to know whether an appeal to consilience is legitimate in any given case, we should look for empirical background assumptions such as the disaparate trace hypothesis, and ask how well supported they are. Which empirical background assumptions are the relevant ones will vary from one context to the next.

So the natural historical attitude can include the idea that consilience, or degree of explanatory unification, gives us some reason to believe (and not merely to accept) a hypothesis. Another consideration that points in this direction has to do with the parity principle. We might appeal to consilience in contexts that have nothing to do with prehistory, and where the inferred events were actually observed by other people. For example, if, when driving down the interstate, you see a twisted guardrail, broken glass, skid marks, etc., you may conclude that the best explanatory unification – i.e. that an accident occurred here – is probably true. It seems inconsistent to allow such inferences in ordinary life but to disallow them in scientific contexts, and only a fairly radical skeptic would disallow such inferences in ordinary life.

Although the constructive empiricist's principled skepticism about non-empirical theoretical virtue goes too far, we still need to approach arguments such as Hoffman and Schrag's argument for the snowball Earth hypothesis with extreme caution. So far, everything I have said about consilience makes it seem as though realism and the natural historical attitude involve what is essentially the same view: Both realists and those who adopt the natural historical attitude will treat consilience as having some evidential weight. At a certain point, though, the two will part ways. If realists claim that explanatory unification is a guide to the real, external past, or to things and events as they existed and occurred independently of us, the proponent of the natural historical attitude will have to demur. The fact that the snowball Earth hypothesis unifies diverse phenomena does give us some reason to think that it is true. But the unifying power of the snowball Earth hypothesis, taken by itself, is irrelevant to the meta-physical question whether global glaciation is something that the scien-tific community discovered, or something that it somehow constructed or brought about.

What's more, conjoining metaphysical claims about mind-dependence or independence to hypotheses about the past will actually tend to make those hypotheses less consilient. The realist who says, "The entire planet froze over, *and* that happened independently of us, our theories, and con-ceptual schemes," is simply tacking on an extra metaphysical claim. This shows that there is an internal tension in the metaphysical realists' view: They want to treat non-empirical theoretical virtue as a guide to truth, but if we do that, historical hypotheses and theories all by themselves will always be more virtuous than those same historical hypotheses and theories conjoined with realist metaphysical claims. From the perspective of the natural historical attitude, the realist metaphysics just looks gratuitous.

8.5 CONSEQUENCES OF THE ASYMMETRIES: SNOWBALL *VS.* SLUSHBALL EARTH

In this chapter, I started out by describing Hoffman and Schrag's appeal to consilience in the snowball Earth debate. I proposed one way (though not the only way) of making the notion of consilience more precise, so as to address the worry that assessments of consilience are simply a matter of "seat-of-the-pants feel." Second, I showed how one can avoid extreme skepticism about such appeals to non-empirical theoretical virtue by treating them as tacit appeals to well-supported background assumptions about historical processes. Perhaps the most interesting question of all, though, is whether the two asymmetries – the asymmetry of manipulability and the role asymmetry of background theories – have any significant consequences for our understanding of the role of consilience in historical research. At first glance, the consequences might even seem paradoxical.

On the one hand, the asymmetries of manipulability and background theories mean that appeals to non-empirical theoretical virtue, such as consilience, explanatory power, and the like, will have a special importance in historical science that they do not have in experimental science. The obvious facts that we cannot intervene in the past, and that historical processes destroy evidence, create special challenges for empirical testing in historical science. This makes appeals to non-empirical virtues such as consilience and unifying power that much more important. The two asymmetries mean that in historical science, a great deal rides on appeals to consilience – indeed more than in any other area of scientific research, although consilience is important in other areas, too.

At this point, the exploration of the significance of the two asymmetries has truly come full circle. Recall the example of Wilson and Carrano's (1999) study of titanosaur hindlimb morphology, from way back in section 1.1. There I described those scientists as "bumping up against" and then transgressing the limits to our knowledge of the past. To begin with, they set out to test a prediction about hindlimb morphology derived from the hypothesis that titanosaurs made wide-gauge trackways. Then, having (arguably) confirmed that hypothesis, they proceeded to indulge in a little speculation: Perhaps the wide-gauge stance of the titanosaurs was an adaptation for a semi-bipedal lifestyle. Rearing up on their hind legs, and using their hind legs and tails to form a kind of tripod, the titanosaurs could have reached foliage that would otherwise have been inaccessible. Wilson and Carrano even list a number of anatomical features that distinguish titanosaurs from other sauropods, and that all seem to make sense under this interpretation. Although they do not quite put it in these terms, they are essentially making an appeal to consilience: The hypothesis that titanosaurs were semi-bipedal may be impossible to test in other ways, but it offers a good explanatory unification of otherwise puzzling features of titanosaur morphology. This seemingly simple case study raises some difficult questions for the epistemology of historical science. How can my earlier suggestion that Wilson and Carrano's hypothesis of semi-bipedalism goes beyond the limits to our knowledge of prehistory – i.e. my suggestion that we cannot reasonably claim to *know* that the titanosaurs were semi-bipedal – square with the view that the degree of consilience of a hypothesis can give us a reason for believing it, or for thinking it is true?

The answer is that there is another reason to be skeptical about appeals to consilience in historical science, and one that has little to do with the more extreme sort of skepticism about such appeals that characterizes constructive empiricism. The problem, well known to epistemologists who work on coherence theories of epistemic justification, is that for any collection of traces, it is often possible to come up with multiple explanatory stories that are about equally consilient. Indeed, this is essentially the same problem that we encountered in the discussion of local underdetermination in chapter 2. In the analysis offered there, local underdetermination problems arise when (among other things) two incompatible hypotheses have equal shares of non-empirical theoretical virtue. There are other potential explanations of the wide-gauge stance of the titanosaurs than the one that Wilson and Carrano offer: perhaps titanosaurs competed for access to mates (or foraging sites, or whatever) by engaging in a side-to-side combat, with each animal leaning into the other, attempting to knock the opponent over, or at least force the other to give ground. Maybe the wide-gauge stance would have increased one's chances of prevailing in such an encounter. Some such story of selection in terms of social behavior could turn out to be just as consilient as Wilson and Carrano's story about semi-bipedal foraging. And what test could scientists ever use to discriminate between them? Or maybe Wilson and Carrano were right about the semi-bipedalism, but wrong about its connection to foraging. Maybe the titanosaurs engaged in elaborate social behaviors in which resting one's chin on the forehead of one's neighbor served as a way of establishing dominance. The problem is not (as constructive empiricists might say) that non-empirical theoretical virtues carry no evidential weight at all; the problem, rather, is that rival hypotheses are often equally virtuous.

This should come as no big surprise, given the incompleteness of the historical record. Given that historical processes typically destroy or hide the evidence, historical researchers will often have to work, as in the snowball Earth case, with a fairly limited set of traces. The fewer the clues, the more room there will be for the free play of the imagination, and the easier it will be to come up with incompatible but equally consilient rival hypotheses.

Hoffman and Schrag initiated the most recent phase of debate about snowball Earth with the publication of their 1998 paper in *Science*. Their appeal to consilience in that paper did carry *some* evidential weight; it did provide some reason for believing the snowball Earth hypothesis. But explanatory unification is always a matter of degree, and in this case it did not provide any conclusive or decisive reason for believing in snowball Earth. In the years since their paper came out, other scientists have developed rival hypotheses – most of which are variations on the snowball Earth theme, but which nevertheless differ from the snowball Earth story in crucial ways. One of these rivals is the high-obliquity hypothesis, with glaciers around the tropical midsection of the planet and warmer ice-free zones at the poles. Another is the "slushball Earth" scenario in which the planet is mostly, but not completely frozen over. I want to suggest, in closing, that the two prevailing models – snowball Earth and slushball Earth – are currently in a state of balance, or equiconsilience. Although this could change in the near future, and may already be changing as scientists discover some anomalies for the "hard" snowball Earth model, it is not inaccurate to say that for the moment, the two models have equal portions of consilience.

The initial impetus for the slushball Earth scenario came from two different sources. First, scientists have long wondered how life could possibly have survived a lengthy snowball Earth episode followed by rapid global warming – the classic "freeze and fry" episode. Even if the earth were totally frozen over, there would be steam vents on the ocean floor, and volcanoes would poke up through the ice here and there. This leaves room for just a few eukaryotic lineages hanging on in marginal environments.

The difficult question is how creatures adapted to life in those extreme environments could have been the ones who also got the Cambrian explosion going.[11] If, however, we suppose that the glacial episodes did not last quite as long, and that a band of open ocean existed in the tropics even though much the rest of the planet was covered in ice, one can tell a more plausible evolutionary story about how events during the neoproterozoic set the stage for the Cambrian explosion. This open water refugium hypothesis also seems to fit better with studies of early metazoan evolution. A number of metazoan lineages show up quite abruptly right at the beginning of the Cambrian, about 565 million years ago. Some of the best explanations to date of the emergence of the earliest metazoan body plans have those metazoan lineages diverging considerably earlier, at or even before the time when the planet was going through the last neoproterozoic glacial episode (Runnegar 2000). It is not easy to imagine how this early divergence of metazoan lineages could have occurred in extreme environments, near steam vents and volcanoes, but it could well have happened in the open water refugium during a slushball Earth episode. The second main source for the slushball Earth hypothesis comes from the numerical experiments carried out by Hyde et al. (2000), and discussed in section 5.6. Those scientists used climate simulations coupled with a model of ocean currents to show that under certain conditions, the ice–albedo feedback effect would not have produced a "hard" snowball Earth. The heat circulated by ocean currents would have preserved an ice-free zone near the tropics.

Not surprisingly, geologists have now begun to look at the evidence that Hoffman and Schrag interpreted as counting in favor of the snowball Earth hypothesis, in order to see if the slushball Earth hypothesis can explain that evidence, too. Much discussion has focused on the strange carbon isotope ratios. Kennedy, Christie-Blick, and Sohl (2001) show how the slushball Earth hypothesis can explain this evidence just as well as the snowball Earth hypothesis can. They focus on terrestrial glaciers, and argue that during a long period of glaciation, large amounts of methane gas from decomposing organic matter would become trapped under the ice in the form of gas hydrates. The phenomenon that both parties want to explain is the higher than expected amount of carbon-12, relative to the amount of carbon-13, in the cap carbonate rocks. Hoffman and Schrag argue that since organisms prefer carbon-12 for photosynthesis, the higher

---

[11] But see McKay (2000), who argues that a thin pack of sea ice (less than 10m thick) would have let enough sunlight through to enable photosynthesis in the tropical oceans.

amounts of carbon-12 suggest that photosynthetic activity in the oceans shut down for some time. On the other hand, Kennedy et al. (2001) argue that the methane gas hydrates trapped under terrestrial glaciers would have contained high amounts of carbon-12 (see also Jacobsen 2001). When the terrestrial glaciers melted, the methane gas was released, and basins and continental shelves were flooded. Carbonate rocks formed in those places would have higher than expected amounts of the carbon-12 isotope. Thus, the snowball and slushball Earth models explain the same phenomenon, but in different ways. The ideal way to discriminate between these two models, each of which has its fair share of consilience, would be to subject them to a novel predictive test. That, as I argued back in chapter 5, is easier said than done, and the issue remains unsettled.

Thus, the effects of the asymmetry of manipulability and the role asymmetry of background theories may seem paradoxical at first. To begin with, they create difficulties for empirical testing that make appeals to nonempirical theoretical virtue, such as consilience, that much more important in historical science. On the other hand, though, they also make the old problem of "multiple coherent stories" that much more acute in historical science: When scientists have incompatible rival models, such as the snowball and the slushball Earth models, which seem about equally consilient, it will be that much more difficult to devise tests to discriminate between them. Here again the asymmetries place historical researchers at an epistemic disadvantage.

Ironically, the best philosophical view of the role that consilience does and should play in historical science turns out to be rather nuanced and complicated. First, radical skepticism about non-empirical theoretical virtue cannot be sustained because of the bootstrapping procedure described earlier. Scientists who began merely by accepting (in van Fraassen's sense) the hypothesis that consilience is a guide to truth would end up making appeals to consilience in the course of their work. Second, the exemplar approach is best way to get clear about what consilience is. The rival analytical approach has so far failed to generate a definition of "consilience" that, in Millgram's words, "the troops can use in the field." Third, the best way to understand how appeals to consilience (and other non-empirical theoretical virtues) can carry evidential weight is to treat them along the lines suggested by Sober, as appeals to empirical background assumptions that can be more or less well supported by experience. The disparate trace hypothesis (or something similar) seems to lie in the background of Hoffman and Schrag's appeal to consilience in the snowball Earth case. Fourth, scientists who adopt the natural historical attitude, as I recommend, should remain cautious about appeals to consilience. Consilience does carry evidential weight. However, someone who adopts the natural historical attitude will remain agnostic about whether consilience is a guide to the truth about mind-independent prehistory, or about events as they happened independently of our present thoughts, theories, and conceptual schemes. Fifth, the asymmetry of manipulability and the role asymmetry of background theories create challenges for empirical testing in historical science, and this means that appeals to consilience do and should have a special prominence in historical research. Historical investigators will often need to appeal to non-empirical theoretical virtue in order to break evidential ties between rival hypotheses. Sixth, the asymmetry of manipulability and the role asymmetry of background theories also imply that the problem of incompatible equiconsilient models (e.g., snowball *vs.* slushball Earth) will be especially common in historical science. This means that the correct attitude to have toward appeals to consilience is one of mitigated skepticism: Yes, appeals to consilience can carry some evidential weight, but it is often possible to dream up equally consilient alternative models, and it can be extremely difficult to carry out the empirical tests needed to discriminate between equiconsilient rivals. There is, in short, nothing simple about appeals to consilience.

Conclusion

Anyone familiar with the scientific realism debate knows that the main players – philosophers such as Richard Boyd, Michael Devitt, and Jarrett Leplin (on the realist side), and Arthur Fine, Larry Laudan, and Bas van Fraassen (on the other side) – have had relatively little to say about the scientific study of prehistory. I have tried to show that as a result of this omission, the realism debate has been skewed in several ways. First, realist arguments meant to establish the possibility of knowledge of unobservables have differential force depending on the kind of unobservables in question. These arguments usually involve some sort of inference to the best explanation. I have tried to show that they give less support to historical than to experimental realism. Second, there are problems concerning the scope of our knowledge of prehistory that do not arise in the context of experimental science. Local underdetermination problems are more common in historical than in experimental science, and novel predictive successes are also fewer and further between. Third, of the two leading varieties of non-realist philosophy of science, one of these (van Fraassen's constructive empiricism) leads to unacceptably radical skepticism when applied to prehistory, whereas Arthur Fine's natural ontological attitude can serve nicely as the basis for a philosophy of historical science – the natural historical attitude. I hope that my arguments for these claims help to move the discussion of scientific realism in a fruitful direction.

In addition, I hope that my approach in this book helps to change the way people think about issues in historical science. Many philosophical discussions of historical science have focused on the distinction between ideographic science (which is oriented toward particular things and events) and nomothetic science (which is oriented toward laws, patterns, and regularities). Instead of taking this distinction as basic, I have tried to shift the focus to kinds of unobservables, where the unobservables

204

are classified according to the facts about them which make them unobservable. This makes it possible to identify some interesting asymmetries between the past and the microphysical (the analogue asymmetry, and especially the asymmetry of manipulability and the role asymmetry of background theories), and to explore their consequences. Even if I turn out to be wrong about some of the details – about the asymmetries and their consequences – I hope that others find this approach to be useful.

The way in which historical science is done is changing, and I hope that this book helps to shed some light on the new developments. In a commentary in *The New York Times* on recent work using genetic evidence to reconstruct hominid evolution, Olivia Judson (2006) writes as though fossils were passé: "[D]ebates have raged about what fossils mean for our understanding of the history of life on Earth, and especially of evolution. No longer. Fossils have become unnecessary to the argument: since we've learned to sequence whole genomes, we've had far more powerful ways to examine the past." Judson is mainly concerned with arguments about evolution, and her claim is that the relative importance of the fossil record, as evidence that evolution has occurred, is diminished by the increasingly useful evidence coming from genetics. We might also see here a case in which relative epistemic disadvantage motivates methodological innovation.

Sometimes the genetic evidence can be difficult to square with the fossil evidence. For example, some scientists have claimed on the basis of fossilized remains that the evolutionary split between our own ancestors and those of chimpanzees occurred around seven million years ago. The oldest skull that has been identified as coming from a hominid (found in Chad in 2002) is about seven million years old. However, a recent study of mutation rates in humans, chimps, and other primates – the study that prompted Judson to argue that fossils are no longer so central to the case for evolution – indicates that the split occurred more recently, between 6.3 and 5.4 million years ago, and also that the split took quite a long time, and that plenty of interbreeding occurred between our ancestors and those of chimps (Patterson et al. 2006). Whether or not the new genetic techniques represent "far more powerful ways to examine the past," Judson overlooks the fact that the new line of evidence also creates a new scientific problem: How do we square what the genetic evidence seems to be telling us with what the fossil evidence seems to say? Notice how similar this is to the methodological problem with which I began the book, back in chapter 1. There I examined the problem of integrating the ichnological evidence (in the form of fossilized sauropod trackways) with the skeletal evidence of sauropod hindlimb morphology. New technology and new methods give rise to a familiar methodological difficulty: How do scientists integrate different lines of evidence?

With a few exceptions, philosophers of science have tended to neglect the kinds of historical research discussed in this book – from the work on snowball Earth, to Wilson and Carrano's study of titanosaur hindlimb morphology, to Raup and Sepkoski's work on the periodicity of extinction events. Although it is difficult to say for sure, this may have something to do with the perception that fields such as paleobiology and geology are somehow less scientific than experimental physics or chemistry. Historically, this perception has gone hand-in-hand with the idiographic/nomothetic distinction. "Real" or "hard" science was supposed to be nomothetic, but paleobiology and geology are perceived to be largely idiographic. I have suggested that paleobiology, geology, and other fields that focus on prehistory really are at an epistemic disadvantage relative to other more experimental fields. However, far from implying that historical science is less good or less scientific than research in other areas, understanding these epistemic disadvantages is crucial to appreciating the recent accomplishments and methodological innovations in historical work, including the use of evidence from genetics in phylogenetic reconstruction. Epistemic disadvantage should not detract from scientific status. Rather, historical researchers deserve credit for coping with epistemic disadvantage.

Finally, I hope to have shown that realism is not the only scientifically respectable view of the study of prehistory. I have argued, in the spirit of Arthur Fine, that we should jettison some of the philosophical theory that is associated with realism – especially theory about the nature of truth and the mind- and theory-independence of the past. The result is a stripped, down, minimalist philosophy of historical science, the natural historical attitude. I have not argued that scientists literally make prehistory. Instead, I have tried to show that realists' attempts to demonstrate the absurdity of the idea that scientists make prehistory all founder. Both realists and their constructivist opponents make metaphysical claims that outrun the available evidence, and we should accordingly suspend judgment about those claims. Fossils and other historical records preserve a great deal of information about the past, but they have nothing to say about metaphysics. Asking whether scientists make prehistory is like asking about the colors of the dinosaurs. Thus, the title of this book serves not as a slogan for any philosophical position, but rather as a reminder of what we cannot know about the past.

References

Acock, M. 1983. "The age of the universe," *Philosophy of Science* 50 (1): 130–145.

Alexander, R. M. 1976. "Estimates of speeds of dinosaurs," *Nature* 261: 129–130.

Allen, C., M. Bekoff and G. Lauder (eds.) 1998. *Nature's purposes: analyses of function and design in biology.* Cambridge, MA: MIT Press.

Alvarez, W. et al. 1980. "Extraterrestrial cause for the cretaceous-tertiary extinction," *Science* 208: 1095–1108.

Anfinrude, P. and F. Schotte 2005. "X-ray fingerprinting of chemical intermediates in solution," *Science* 309 (5738): 1192–1193.

Bakker, R. T. 1986. *The dinosaur heresies.* New York: Kensington Publishing.

Barlow, C. 2000. *The ghosts of evolution: Nonsensical fruit, missing partners, and other ecological anachronisms.* New York: Basic Books.

Bell, G. L., Jr. 1997. "Introduction [to the Mosasauridae]," in J. M. Callaway and E. L. Nicholls (eds.), *Ancient marine reptiles.* New York: Academic Press, 281–292.

Ben-Menahem, Y. 1997. "Historical contingency," *Ratio* 10 (2): 991–1007.

Benton, M. J. 1995. "Diversification and extinction in the history of life," *Science* 268 (5207): 52–58.

1997. "Models for the diversification of life," *Trends in Ecology and Evolution* 12 (12): 490–495.

1999. "The history of life: Large databases in palaeontology," in D. A. T. Harper (ed.), *Numerical palaeobiology: Computer-based modeling and analysis of fossils and their distribution.* New York: John Wiley and Sons, pp. 249–283.

Blackburn, S. 2002. "Realism: Deconstructing the debate," *Ratio* 15 (2): 111–133.

2005. *Truth: A guide.* Cambridge, MA: Harvard University Press.

Boyd, R. 1985. "Lex orandi est lex credendi," in P. M. Churchland and C. A. Hooker (eds.), *Images of science.* Chicago: University of Chicago Press, pp. 3–34.

1984. "The current status of scientific realism," in J. Leplin (ed.), *Scientific realism.* Berkeley, CA: University of California Press, pp. 41–82.

1990. "Realism, approximate truth, and method," *Minnesota Studies in the Philosophy of Science* 14: 355–391.

Buller, D. (ed.) 1999. *Function, selection, and design.* Albany, NY: SUNY Press.

References

Carman, C. 2005. "The electrons of the dinosaurs and the center of the earth," *Studies in History and Philosophy of Science* 36 (1): 171–174.

Cleal, C. J., and B. A. Thomas 1999. *Plant fossils: The history of land vegetation.* Rochester, NY: The Boydell Press.

Cleland, C. 2002. "Methodological and epistemic differences between historical science and experimental science," *Philosophy of Science* 69 (3): 474–496.

Coles, J. 1973. *Archaeology by experiment.* London: Hutchinson and Company.

Collins, D. 1996. "The 'evolution' of Anomalocaris and its classification in the Arthropod class Dinocarida and order Radiodonta," *Journal of Paleontology* 70: 280–293.

Cuddington, K., and M. Ruse 2004. "Biodiversity, Darwin, and the fossil record," in M. Oksanen and J. Pietarinen (eds.), *Philosophy and biodiversity.* Cambridge: Cambridge University Press, pp. 101–118.

Devitt, M. 1991. *Realism and truth,* second edition. Princeton, NJ: Princeton University Press.

Donnadieu, Y., F. Fluteau, G. Ramstein, et al. 2003. "Is there a conflict between the Neoproterozoic glacial deposits and the snowball Earth interpretation: an improved understanding with numerical modeling," *Earth and Planetary Science Letters* 208: 101–112.

Dummett, M. 1979. "The reality of the past," in *Truth and other enigmas.* Cambridge, MA: Harvard University Press, pp. 358–374.

  2003. "Truth and the past," *Journal of Philosophy,* 100: 5–53.

Evans, D. A. D. 2000. "Stratigraphic, geochronological, and paleomagnetic constraints upon the neoproterozoic climate paradox," *American Journal of Science* 300: 347–433.

Farlow, J. O. 1992. "Sauropod tracks and trackmakers: integrating the ichnological and skeletal records," *Zubia* 10: 89–138.

Farlow, J. O. and M. G. Lockley 1989. "Roland T. Bird, dinosaur tracker: An appreciation," in D. D. Gillette and M. G. Lockley (eds.), *Dinosaur tracks and traces.* Cambridge: Cambridge University Press, pp. 33–36.

Feinberg, G., S. Lavine, and D. Albert 1992. "Knowledge of the past and the future," *The Journal of Philosophy* 89: 607–642.

Fine, A. 1984. "The natural ontological attitude," in J. Leplin (ed.), *Scientific realism.* Berkeley, CA: University of California Press, 83–107.

  1986. "Unnatural attitudes: Realist and instrumentalist attachments to science," *Mind* 95: 149–179.

  1996. *The shaky game: Einstein, realism, and the quantum theory,* second edition. Chicago, IL: University of Chicago Press.

Fisher, P. E. et al. 2000. "Cardiovascular evidence for an intermediate or higher metabolic rate in an ornithiscian dinosaur," *Science* 288: 503–505.

Friedman, M. 1974. "Explanation and scientific understanding," *Journal of Philosophy,* 71: 5–10.

Gallie, W. B. 1959. "Explanations in history and the genetic sciences," in P. Gardiner (ed.), *Theories of history.* Glencoe, IL: The Free Press.

Gero, J. M. 1989. "Producing prehistory, controlling the past: the case of New England beehives," in V. Pinsky and A. Wylie (eds.), *Critical traditions in contemporary archaeology: Essays in the philosophy, history, and socio-politics of archaeology.* Albuquerque, NM: University of New Mexico Press, pp. 96–103.

Gottelli, D., C. Silleri-Zubiri, G. D. Appelbaum et al. 1994. "Molecular genetics of the most endangered canid: the Ethiopian Wolf, *Canis simensis,*" *Molecular Ecology* 3: 301–312.

Gould, S. J. 1980. "The promise of paleobiology as a nomothetic, evolutionary discipline," *Paleobiology* 6 (1): 96–118.

1987. *Time's arrow, time's cycle: Myth and metaphor in the discovery of geological time.* Cambridge, MA: Harvard University Press.

1989. *Wonderful life: The Burgess Shale and the nature of history.* New York: W.W. Norton.

1991. *Bully for Brontosaurus.* New York: W. W. Norton.

2002. *The structure of evolutionary theory.* Cambridge, MA: Harvard University Press.

Gould, S. J., and R. Lewontin 1979. "The spandrels of San Marco and the Panglossian paradigm: A critique of the adaptationist programme," *Proceedings of the Royal Society* B205: 581–598.

Grantham, T. 1999. "Explanatory pluralism in paleobiology," *Philosophy of Science* 66 (supp.): S223-S236.

Greene, B. 1999. *The elegant universe: Superstrings, hidden dimensions and the quest for the ultimate theory.* New York: W. W. Norton and Company.

Hacking, I. 1983. *Representing and intervening: Introductory topics in the philosophy of natural science.* Cambridge: Cambridge University Press.

1999. *The social construction of what?* Cambridge, MA: Harvard University Press.

Hanen, M. and J. Kelley 1989. "Inference to the best explanation in archaeology," in V. Pinsky and A. Wylie (eds.), *Critical traditions in contemporary archaeology: Essays in the philosophy, history, and socio-politics of archaeology.* Albuquerque, NM: University of New Mexico Press, pp. 14–17.

Harré, R. 1986. *Varieties of realism: A rationale for the natural sciences.* Oxford: Blackwell.

1996. "From observability to manipulability: Extending the inductive arguments for realism," *Synthese* 108 (2): 137–155.

Hempel, C. G. 1965. *Aspects of scientific explanation, and other essays in the philosophy of science.* New York: The Free Press.

1966. *Philosophy of natural science.* Englewood Cliffs, NJ: Prentice-Hall.

Hitchcock, E. 1858/1974. *A Report on the sandstone of the Connecticut valley, especially its fossil footmarks.* New York: Arno Press.

Hoffman, P. F., A. J. Kaufman, G. P. Halverson, and D. P. Schrag. 1998. "A neoproterozoic snowball Earth," *Science* 281 (5381): 1342–1346.

Hoffman, P. F. and D. P. Schrag 2000. "Snowball Earth," *Scientific American* 282 (1): 68–75.

Hopson, J. A. 1975. "The evolution of cranial display structures in hadrosaurine dinosaurs," *Paleobiology* 1: 21–43.

Horwich, P. 1982. *Probability and evidence.* Cambridge: Cambridge University Press.

1987. *Asymmetries in time.* Cambridge, MA: MIT Press.

1991. "On the nature and norms of theoretical commitment," *Philosophy of Science* 38 (51): 1–14.

Hughes, N. 1999. "Statistical and imaging methods applied to deformed fossils," in D. A. T. Harper, (ed.), *Numerical paleobiology: Computer-based modeling and analysis of fossils and their distributions.* New York: John Wiley and Sons, pp. 126–155.

Hull, D. L. 1975. "Central subjects and historical narratives," *History and Theory* 14: 253–274.

Huss, J. E. 2004. "Experimental reasoning in non-experimental science: Case studies from paleobiology," unpublished PhD thesis, University of Chicago. UMI number 3125619.

Hyde, W. T., T. J. Crowley, S. K. Baum, and W. R. Peltier 2000. "Neoproterozoic 'snowball Earth' simulations with a coupled climate/ice sheet model," *Nature* 405 (6785): 425–429.

Jacobsen, S. B. 2001. "Gas hydrates and deglaciations," *Nature* 412 (6848): 691–693.

Jaffe, M. 2000. *The gilded dinosaur.* New York: Crown Publishers.

Janzen, D. H. and P. S. Martin 1982. "Neotropical anachronisms: The fruits the Gomphotheres ate," *Science* 215: 19–27.

Jenkins, G. S. 2000. "The 'Snowball Earth' and Precambrian climate," *Science* 288: 975–976.

Judson, O. 2006. "Affairs to remember," *The New York Times* (May 28), section 4, p. 10.

Kasting and Caldeira 1992. "Susceptibility of the early earth to irreversible glaciation caused by carbon-dioxide clouds," *Nature* 359: 226–228.

Kemp, T. S. 1999. *Fossils and evolution.* Oxford: Oxford University Press.

Kennedy, M. J., N. Christie-Blick, and L. E. Sohl 2001. "Are proterozoic cap carbonates and isotopic excursions a record of gas hydrate destabilization following Earth's coldest intervals?," *Geology* 29 (5): 446–446.

Kerr, R. 2000. "An appealing snowball Earth that's still hard to swallow," *Science* 287 (5459): 1734–1736.

Kirkham, R. L. 1992. *Theories of truth: A critical introduction.* Cambridge, MA: MIT Press.

Kirschvink, J. L. 1992. "Late proterozoic low-latitude global glaciation: The snowball Earth," in J. W. Schopf and C. Klein (eds.), *The proterozoic biosphere: A multidisciplinary approach.* New York: Cambridge University Press.

Kirschvink, J. L., E. J. Gaidos, L. E. Bertani et al. 2000. "Paleoproterozoic snowball Earth: Extreme climatic and geochemical global change and its biological consequences," *Proceedings of the National Academy of Sciences* 97 (4): 1400–1405.

Kitcher, P. 1989. "Explanatory unification and the causal structure of the world," *Minnesota Studies in the Philosophy of Science.* 13: 410–450.

1993. *The advancement of science: science without legend, objectivity without illusions.* Oxford: Oxford University Press.

Kleinhans, M. G., C. J. J. Buskes, and H. W. de Regt 2005. "Terra Incognita: Explanation and reduction in earth science," *International Studies in Philosophy of Science* 19 (3): 289–317.

Kosso, P. 2001. *Knowing the past: Philosophical issues of history and archaeology.* Amherst, NY: Humanity Books.

Kuhn, T. 1996. *The structure of scientific revolutions.* Chicago: University of Chicago Press.

Kukla, A. 1994. "Non-empirical theoretical virtues and the argument from under-determination," *Erkenntnis* 41: 157–170.

1996. "Does every theory have empirically equivalent rivals?," *Erkenntnis* 44: 137–166.

2000. *Social constructivism and the philosophy of science.* London: Routledge.

Ladyman, J. 1999. "Review of Jarrett Leplin, *A novel defense of scientific realism,*" *British Journal for the Philosophy of Science* 50: 181–188.

Laudan, L. 1971. "William Whewell on the consilience of inductions," *Monist* 55: 368–391.

1981/1996. "A confutation of convergent realism," in D. Papineau (ed.), *The philosophy of science.* Oxford: Oxford University Press, pp. 107–138.

Laudan, L. and J. Leplin 1991. "Empirical equivalence and underdetermination," *Journal of Philosophy* 88: 449–472.

Latour, B. and S. Woolgar 1986. *Laboratory life: The construction of scientific facts.* Princeton, NJ: Princeton University Press.

Leather, J., P. A. Allen, M. D. Brasier, and A. Cozzi 2002. "Neoproterozoic snow-ball Earth under scrutiny: Evidence from the Fiq glaciation of Oman," *Geology* 30 (10): 891–894.

Leplin, J. 1997. *A novel defense of scientific realism.* Oxford: Oxford University Press.

Lewis, D. 1979. "Counterfactual dependence and time's arrow," *Nous* 13: 455–476.

Lipton, P. 1991. *Inference to the best explanation.* London: Routledge.

1993. "Is the best good enough?," *Proceedings of the Aristotelian Society* 93: 89–104.

Lockley, M. G. et al. 1994. "The distribution of sauropod tracks and trackmakers," *Gaia* 10: 233–248.

Lynch, M. P. 2004. *True to life: Why truth matters.* Cambridge, MA: MIT Press.

McGrew, T. 2003. "Confirmation, heuristics, and explanatory reasoning," *British Journal for the Philosophy of Science* 54: 553–567.

McKay, C. P. 2000. "Thickness of tropical ice and photosynthesis on a snowball Earth," *Geophysical Research Letters* 27 (14): 2153–2156.

McMullin, E. 1984. "A case for scientific realism," in J. Leplin (ed.), *Scientific realism.* Berkeley, CA: University of California Press, pp. 8–40.

Martin, L. D. and B. M. Rothschild 1987. "Avascular necrosis: Occurrence in diving cretaceous mosasaurs," *Science* 236: 75–77.

Martin, L. D. and B. M. Rothschild 1989. "Paleopathology and diving mosasaurs," *American Scientist* 77: 460–466.

Massare, J. 1988. "Swimming capabilities of Mesozoic marine reptiles: implications for method of predation," *Paleobiology* 14 (2): 187–205.

Maxwell, G. 1962/1999. "The ontological status of theoretical entities," in H. Feigl and G. Maxwell (eds.), *Minnesota Studies in The Philosophy of Science,* vol. III.

211

Mayo, D. 1991. "Novel evidence and severe tests," *Philosophy of Science* 58: 523–553.

—— 1997. "Severe tests, arguing from error, and methodological underdetermination," *Philosophical Studies* 86: 243–266.

Melnyk, A. 1997. "How to keep the "physical' in physicalism," *Journal of Philosophy* 94 (12): 622–637.

Millgram, E. 2000. "Coherence: The price of the ticket," *Journal of Philosophy* 97 (2): 82–93.

Mitchell, W. J. T. 1998. *The last dinosaur book: The life and times of a cultural icon.* Chicago: University of Chicago Press.

Musgrave, A. 1989. "NOA's Ark – fine for realism," *Philosophical Quarterly* 39: 383–398.

—— 1999. "Idealism and antirealism," in R. Klee (ed.), *Scientific inquiry.* Oxford: Oxford University Press, 344–352.

Nagel, E. 1979. *The structure of science: Problems in the logic of scientific explanation.* Indianapolis, IN: Hackett Publications.

Newton-Smith, W. H. 1987. "Realism and inference to the best explanation," *Fundamenta Scientiae* 7: 305–316.

Nojima, S., C. Schal, F. X. Webster et al. 2005. "Identification of the sex pheromone of the German cockroach, *Blattella germanica*," *Science* 307 (5712): 1104–1106.

Nola, R. 2002. "Realism through manipulation, and by hypothesis," in S. Clarke and T. D. Lyons (eds.), *Recent themes in the philosophy of science: scientific realism and commonsense.* Dordrecht: Kluwer Academic Publishers, pp. 1–24.

Noonan, J. P., M. Hofreiter, D. Smith et al. 2005. "Genomic sequencing of pleistocene cave bears," *Science* 309 (5734): 597–600.

Norman, D. 1988. *The prehistoric world of the dinosaur.* New York: Gallery Books.

Padian, K. and P. E. Olsen 1984. "The track of *Pteraichnus*: not Pterosaurian, but Crocodilian," *Journal of Paleontology* 58: 178–184.

—— 1989. "Ratite footprints and the stance and gait of mesozoic therapods," in D. D. Gillette and M. Lockley (eds.), *Dinosaur tracks and traces.* Cambridge: Cambridge University Press, pp. 231–242.

Parsons, K. M. 2001. *Drawing out leviathan: dinosaurs and the science wars.* Bloomington, IN: Indiana University Press.

Patterson, C. and A. B. Smith 1987. "Is periodicity of mass extinctions a taxonomic artifact?," *Nature* 330: 248–251.

—— 1989. "Periodocity in extinction: the role of systematics," *Ecology* 70: 802–811.

Patterson, N., D. J. Richter, S. Gnerre et al. 2006. "Genetic evidence for complex speciation of humans and chimpanzees," *Nature* (advance online publication 17 May). Available online at www.nature.com/nature/journal/vaop/ncurrent/pdf/nature04789.pdf. Last accessed June 4, 2006.

Post, J. F. 1987. *The faces of existence: An essay in nonreductive metaphysics.* Ithaca, NY: Cornell University Press.

—— 1996. "The foundationalism in irrealism, and the immorality," *Journal of Philosophical Research, 21*, 1–14.

Psillos, S. 1999. *Scientific realism: How science tracks truth.* London: Routledge.

Putnam, H. 1978. *Meaning and the moral sciences*. London: Routledge and Kegan Paul.

1981. *Reason, truth, and history*. Cambridge: Cambridge University Press.

1984. "What is realism?," in J. Leplin (ed.), *Scientific realism*. Berkeley: University of California Press, pp. 140–153.

Quine, W. V. 1975. "On empirically equivalent systems of the world," *Erkenntnis* 9: 313–328.

Raup, D. M., S. J. Gould, T. J. M. Schopf, and D. S. Simberloff 1973. "Stochastic models of phylogeny and the evolution of diversity," *Journal of Geology* 81: 525–542.

Raup, D. M. and S. J. Gould 1974. "Stochastic simulation and evolution of morphology – Towards a nomothetic paleobiology," *Systematic Zoology* 23 (2): 305–322.

Raup, D. M. and J. J. Sepkoski, Jr. 1984. "Periodicity of extinctions in the geologic past," *Proceedings of the National Academy of Sciences* 81: 801–805.

1986. "Periodic extinctions of families and genera," *Science* 231: 833–836.

Rosenberg, A. 1996. "A field guide to recent species of naturalism," *British Journal for the Philosophy of Science* 47 (1): 1–29.

Rouse, J. 1996. *Engaging science: How to understand its practices philosophically*. Ithaca, NY: Cornell University Press.

Rowe, T., E. F. McBride, and P. C. Sereno 2001. "Dinosaur with a heart of stone," *Science* 291: 783a.

Runnegar, B. 2000. "Loophole for snowball Earth," *Nature* 405 (6785): 403–404.

Ruse, M. 1999. *Mystery of mysteries: Is evolution a social construction?* Cambridge, MA: Harvard University Press.

Russell, B. 1921. *The analysis of mind*. London: Allen and Unwin.

Salmon, M. H. 1982. *Philosophy and archaeology*. New York: Academic Press.

Sato, T., Y. Cheng, X. Wu et al. 2005. "A pair of shelled eggs inside a female dinosaur," *Science* 308 (5720): 375.

Savitt, S. F., 1990. "Epistemological time asymmetry," *PSA: Proceedings of the Biennial Meeting of the Philosophy of Science Association* 1: 317–324.

Schweitzer, M. H., J. L. Wittmeyer, J. R. Horner, and J. K. Toporski 2005. "Soft-tissue vessels and cellular preservation in *Tyrannosaurus rex*," *Science* 307 (5717):1952–1955.

Sepkoski, J. J., Jr. 1978. "A kinetic model of Phanerozoic taxonomic diversity. I. Analysis of marine orders," *Paleobiology* 4: 223–251.

1979. "A kinetic model of Phanerozoic taxonomic diversity. II. Early Paleozoic families and multiple equilibria," *Paleobiology* 5: 222–252.

1982. *A compendium of fossil marine families*. Milwaukee Public Museum Contributions in Biology and Geology 51.

1984. "A kinetic model of Phanerozoic taxonomic diversity. III. Post-paleozoic families and mass extinctions," *Paleobiology* 10: 246–267.

Sepkoski, J. J., Jr. and D. C. Kendrick 1993. "Numerical experiments with model monophyletic and paraphyletic taxa," *Paleobiology* 19: 168–184.

Sheldon, A. 1997. "Ecological implications of mosasaur bone microstructure," in J. M. Callaway and E. L. Nicholls, (eds.), *Ancient Marine Reptiles*. New York: Academic Press.

Smart, J. J. C. 1963. *Philosophy and scientific realism*. London: Routledge.

Sober, E. 1988. *Reconstructing the past: Parsimony, evolution, and inference*. Cambridge, MA: MIT Press.

Stanford, P. K. 2000. "An antirealist explanation of the success of science," *Philosophy of Science* 67 (2): 266–284.

2001. "Refusing the Devil's bargain: What kind of underdetermination should we take seriously?," *Philosophy of Science* 68(Supplement): S1-S12.

Steinbok, R. T. 1989. "Ichnology of the Connecticut Valley: A vignette of American science in the early nineteenth century," in D. D. Gillette and M. G. Lockley (eds.), *Dinosaur tracks and traces*. Cambridge: Cambridge University Press, pp. 27–32.

Taylor, M. A. 1994. "Stone, bone, and blubber? Buoyancy control strategies in aquatic tetrapods," in L. Maddock, Q. Bone, and J. M. V. Rayner (eds.), *Mechanisms and Physiology of Animal Swimming*. New York: Cambridge University Press, 151–161.

Thagard, P. 1991. "The dinosaur debate: explanatory coherence and the problem of competing hypotheses," in J. Pollock and R. Cummins, (eds.), *Philosophy and AI: essays at the interface*. Cambridge, MA: MIT Press, pp. 279–300.

1992. *Conceptual revolutions*. Princeton, NJ: Princeton University Press.

2000. *Coherence in thought and action*. Cambridge, MA: MIT Press.

Thagard, P. and K. Verbeurgt 1998. "Coherence and constraint satisfaction," *Cognitive Science* 22: 1–24.

Thulborn, R. A. 1989. "The gaits of dinosaurs," in D. D. Gillette and M. G. Lockley (eds.), *Dinosaur tracks and traces*. Cambridge: Cambridge University Press, pp. 39–50.

1990, *Dinosaur tracks*. New York: Chapman and Hall.

Tucker, A. 1998. "Unique events: the underdetermination of explanation," *Erkenntnis* 48 (1): 59–80.

2004. *Our knowledge of the past*. Cambridge: Cambridge University Press.

Turner, D. 2000. "The functions of fossils: Inference and explanation in functional morphology," *Studies in History and Philosophy of Biological and Biomedical Science* 31: 137–164.

2004. "The past vs. the tiny: Historical science and the abductive arguments for realism," *Studies in History and Philosophy of Science* 35: 1–17.

2005a. "Local underdetermination in historical science," *Philosophy of Science* 72: 209–230.

2005b. "Misleading observable analogues in paleontology," *Studies in History and Philosophy of Science* 36: 175–183.

van Fraassen, B. C. 1980. *The scientific image*. Oxford: Oxford University Press.

1983. "Glymour on evidence and explanation," in J. Earman (ed.), *Testing scientific theories: Minnesota studies in the philosophy of science*, vol. 10. Minneapolis: University of Minnesota Press, pp. 165–176.

1985. "Empiricism in philosophy of science," in P. M. Churchland and C. A. Hooker, (eds.), *Images of science: Essays on realism and empiricism*. Chicago, IL: University of Chicago Press, pp. 245–308.

1989. *Laws and symmetry*. Oxford: Oxford University Press.

2002. *The empirical stance*. New Haven: Yale University Press.

Weishampel, D. B. 1981. "Acoustic analysis of potential vocalization in lambeosaurine dinosaurs," *Paleobiology* 7: 252–261.

1997. "Dinosaurian cacophony: Inferring function in extinct organisms," *Bioscience* 47: 150–159.

Whiteaves, J. 1892. "Description of a new genus of phyllocarid Crustacea from the Middle Cambrian of Mount Stephen, BC," *Canadian Record of Science* 5: 205–208.

Whittington, H. B. and D. E. G. Briggs 1985. "The largest Cambrian animal, Anomalocaris, Burgess Shale, British Columbia," *Philosophical Transactions of the Royal Society of London* B309: 569–609.

Wilson, J. A. and M. T. Carrano 1999. "Titanosaurs and the origin of "wide-gauge' trackways: a biomechanical and systematic perspective on sauropod locomotion," *Paleobiology* 25 (2), 252–267.

Wittgenstein, L. 1969. *On Certainty*. New York: Harper and Rowe.

Wylie, A. 1995. "Unification and convergence in archaeological explanation: the agricultural 'wave of advance' and the origins of Indo-European languages," *Southern Journal of Philosophy* 34 (supp): 1–30.

2002. *Thinking from things: Essays in the philosophy of archaeology*. Berkeley, CA: University of California Press.

Zack, N. 2002. *Philosophy of science and race*. London: Routledge.

Index

abduction, 68, 105, 193, 204
  argument from novel predictive
    success, 106
  and circularity, 78
  argument for historical realism, 80
  argument for scientific realism, 32,
    69, 78
  and consilience, 192
  criticisms of, 79
  skepticism about, 79
acceptance (of a theory), 173, 192, 193
  epistemic dimension of, 194
Acock, M., 48
ad hocness, 105
adaptationism, 51
agnosticism, 155, 156, 159
  with respect to metaphysical claims,
    130, 160, 176, 178
  with respect to the nature of truth,
    176, 177, 178
Albert, D., 18
Alexander, R. M., 7
Allen, C., 14
Alvarez, L., 6, 40
analogue asymmetry, 86, 87, 94, 100,
    205
  consequences of, 95
  defined, 86
Anomalocaris, 88
Apatosaurus, 11
archaeology, 2, 63, 72, 87, 112, 138, 180
Archaeopteryx, 120

argument from historical narrative,
    109
argument from the bad lot, 79
asymmetry of background theories, 2,
    10, 26, 33, 36, 96, 101, 164, 174,
    178, 179, 180, 205
  and underdetermination, 37
  consequences for the argument
    from novel predictive success, 106
  consequences of, 202
  and consilience, 198
  and novel prediction, 115
asymmetry of manipulability, 2, 10, 26,
    33, 36, 96, 99, 101, 164, 174, 178,
    179, 180, 205
  consequences for the argument
    from novel predictive success, 106
  consequences for the pessimistic
    induction, 100
  consequences of, 61, 202
  and consilience, 198
  defined, 24
  and historical hypo-realism, 68
  and novel prediction, 115
  and numerical modeling, 125, 129
asymmetry of overdetermination, 37,
    38, 39, 40, 41, 42, 43, 44, 45, 60,
    104, 195
asymmetry of recording and
    precording systems, 22, 26
Australopithecus afarensis, 123
avascular necrosis, 111

and warranted assertibility, 138,
   160
Tucker, A., 7, 11, 21, 109
Turner, D., 14, 88
Tylosaurus, 111, 112
Tyrannosaurus, 56

underdetermination, 31, 37, 44, 53,
   155, 178, 184
 and the analogue asymmetry, 86
 global, 48, 49, 55, 171
 local, 37, 43, 45, 46, 47, 48, 49, 53, 55,
   56, 57, 59, 86, 109, 199, 204
unifying role, 70, 79
uniqueness condition, 101, 108, 111,
   120
 and smoking guns, 39
unobservable entities, 23
 kinds of, 34, 35, 36, 61, 63, 65, 67, 204
 producing *vs.* unifying role, 70, 71,
   81

vagueness, 66
Van Fraassen, B.C., 4, 66, 75, 79, 151,
   159, 163, 167, 168, 169, 172, 173,
   174, 178, 194, 204

on abduction, 69
on the belief/acceptance distinction,
   167, 168
on empirical adequacy, 167
on the observable/unobservable
   distinction, 170, 171
Verbeurgt, K., 190
 on coherence, 191
verificationism, 167
verisimilitude, 1

Wegener, A., 82, 181
weighted coherence theory, 190
Weishampel, D., 88
Whewell, W., 182
Whiteaves, J., 88
Whittington, H., 88
Wilson, J., 13, 14, 25, 120, 198, 199,
   206
Windelband, W., 7
Wittgenstein, L., 48
Wittmeyer, J., 56
Woolgar, S., 141, 142, 144, 145
Wylie, A., 2, 180

Zack, N., 137

Milton Keynes UK
Ingram Content Group UK Ltd.
UKHW041519181024
449640UK00003B/18